Lecture Notes in Mathematics 830

Lecture Notes in Mathematics 830

Editors:
J.-M. Morel, Cachan
F. Takens, Groningen
B. Teissier, Paris

J.A. Green

Polynomial Representations of GL_n

2nd corrected and augmented edition

with an Appendix on Schensted Correspondence and Littelmann Paths

by K. Erdmann, J.A. Green and M. Schocker

Springer

Author and co-authors for the appendix

James A. Green
19 Long Close
Oxford OX2 9SG
United Kingdom
e-mail: james.green@maths.ox.ac.uk

Manfred Schocker
Department of Mathematics
University of Wales Swansea
Singleton Park, Swansea SA2 8PP
United Kingdom
e-mail: m.schocker@swansea.ac.uk

Karin Erdmann
Mathematical Institute
University of Oxford
24-29 St Giles
Oxford OX1 3LB
United Kingdom
e-mail: erdmann@maths.ox.ac.uk

Library of Congress Control Number: 2006934862

Mathematics Subject Classification (2000): Primary: 20C30, 20G05, 20G15, 16S50, 17B99, 05E10

ISSN print edition: 0075-8434
ISSN electronic edition: 1617-9692
ISBN 3-540-46944-3 Springer Berlin Heidelberg New York
ISBN 978-3-540-46944-5 Springer Berlin Heidelberg New York

DOI 10.1007/3-540-46944-3

Typesetting by the authors using a Springer LaTeX package
Cover design: WMXDesign GmbH, Heidelberg

Printed on acid-free paper SPIN: 11008118 VA41/3100/SPi 5 4 3 2 1 0

Preface to the second edition

This second edition of "Polynomial representations of $\mathsf{GL}_n(K)$" consists of two parts. The first part is a corrected version of the original text, formatted in LaTeX, and retaining the original numbering of sections, equations, etc. The second is an Appendix, which is largely independent of the first part, but which leads to an algebra $L(n, r)$, defined by P. Littelmann, which is analogous to the Schur algebra $S(n, r)$. It is hoped that, in the future, there will be a structure theory of $L(n, r)$ rather like that which underlies the construction of Kac-Moody Lie algebras.

We use two operators which act on "words". The first of these is due to C. Schensted (1961). The second is due to Littelmann, and goes back to a 1938 paper by G. de B. Robinson on the representations of a finite symmetric group. Littelmann's operators form the basis of his elegant and powerful "path model" of the representation theory of classical groups. In our Appendix we use Littelmann's theory only in its simplest case, i.e. for GL_n.

Essential to my plan was to establish two basic facts connecting the operations of Schensted and Littelmann. To these "facts", or rather conjectures, I gave the names Theorem A and Proposition B. Many examples suggested that these conjectures are true, and not particularly deep. But I could not prove either of them.

This work was therefore stalled, until I sought the help of my colleagues Karin Erdmann and Manfred Schocker. They accepted the challenge, and within a few weeks produced proofs of both conjectures. Their proofs constitute the heart of the Appendix, and make it possible to begin a comparison of the Littelmann algebra $L(n, r)$ with the Schur algebra $S(n, r)$. Karin and Manfred have made this Appendix possible, and have written large parts of the text. It has been a happy experience for me to work with them.

A few weeks before the final manuscript of the Appendix was ready, we heard that A. Lascoux, B. Leclerc and J.-Y. Thibon have published a work on "The plactic monoid", which contains results equivalent to Theorem A and Proposition B. Their methods are rather different from ours, and they prove also many important facts which do not come into our Appendix. We give a brief summary of this work in §D.11.

Oxford, August 2006 Sandy (J. A.) Green

Contents

1

Introduction

Issai Schur determined the polynomial representations of the complex general linear group $\mathsf{GL}_n(\mathbb{C})$ in his doctoral dissertation [47], published in 1901. This remarkable work contained many very original ideas, developed with superb algebraic skill. Schur showed that these representations are completely reducible, that each irreducible one is "homogeneous" of some degree $r \geq 0$ (see 2.2), and that the equivalence types of irreducible polynomial representations of $\mathsf{GL}_n(\mathbb{C})$, of fixed homogeneous degree r, are in one-one correspondence with the partitions $\lambda = (\lambda_1, \ldots, \lambda_n)$ of r into not more than n parts. Moreover Schur showed that the character of an irreducible representation of type λ is given by a certain symmetric function S_λ in n variables (since described as "Schur function"; see 3.5). An essential part of Schur's technique was to set up a correspondence between representations of $\mathsf{GL}_n(\mathbb{C})$ of fixed homogeneous degree r, and representations of the finite symmetric group $G(r)$ on r symbols, and through this correspondence to apply G. Frobenius' discovery of the characters of $G(r)$ (see [17]).

This pioneering achievement of Schur was one of the main inspirations for Hermann Weyl's monumental researches on the representation theory of semi-simple Lie groups [54]. Of course Weyl's methods, based on the representation theory of the Lie algebra of the Lie group Γ, and the possibility of integrating over a compact form of Γ, were very different from the purely algebraic methods of Schur's dissertation; in particular Weyl's general theory contained nothing to correspond to the symmetric group $G(r)$. In 1927 Schur published another paper [48] on $\mathsf{GL}_n(\mathbb{C})$, which has deservedly become a classic. In this he exploited the "dual" actions of $\mathsf{GL}_n(\mathbb{C})$ and $G(r)$ on r^{th} tensor space $E^{\otimes r}$ (see 2.6) to rederive all the results of his 1901 dissertation in a new and very economical way. Weyl publicized the method of Schur's 1927 paper, with its attractive use of the "double centralizer property", in his influential book "The Classical Groups" [55]. In fact the exposition in Chapters 3B and 4 of that book has become a standard treatment of polynomial representations of $\mathsf{GL}_n(\mathbb{C})$ (and, incidentally, of Alfred Young's representation theory of the symmetric group $G(r)$), and perhaps this explains the comparative neglect of

Schur's work of 1901. I think this neglect is a pity, because the methods of this earlier work are in some ways very much in keeping with the present-day ideas on representations of algebraic groups. It is the purpose of these lectures to give some accounts, in part based on the ideas of Schur's 1901 dissertation, of the polynomial representations of the general linear groups $\mathsf{GL}_n(K)$, where K is an infinite field of arbitrary characteristic.

Our treatment will be "elementary" in the sense that we shall not use algebraic group theory in our main discussion. But it might be interesting to indicate here some general ideas from the representation theory of algebraic groups (or algebraic semigroups, since the group inverse is not important in this context), which are relevant to our work.

Let Γ be any semigroup (i.e. Γ is a set, equipped with an associative multiplication) with identity 1_Γ, and let K be any field. A *representation* τ of Γ on a K-space V (i.e. a vector space over K) is a map $\tau : \Gamma \to \mathsf{End}_K(V)$ which satisfies $\tau(gg') = \tau(g)\tau(g')$, $\tau(1_\Gamma) = \mathbb{I}_V$, for all $g, g' \in \Gamma$. (For any set V, we denote by $\mathbb{I}_V$ the identity map on V.) We can extend τ linearly to give a map of K-algebras $\tau : K\Gamma \to \mathsf{End}_K(V)$; here $K\Gamma$ is the *semigroup-algebra* of Γ over K, whose elements are all formal linear combinations

$$\kappa = \sum_{g\in\Gamma} \kappa_g g, \quad \kappa_g \in K,$$

whose support $\mathsf{supp}\,\kappa = \{\, g \in \Gamma : \kappa_g \neq 0 \,\}$ is finite. We can make $K\Gamma$ act on V by $\kappa v = \tau(\kappa)(v)$ ($\kappa \in K\Gamma$, $v \in V$), and thereby get a left $K\Gamma$-module, denoted (V, τ), or simply V. A $K\Gamma$*-map* between such $K\Gamma$-modules (V, τ), (V', τ') is, by definition, a K-map $f : V \to V'$ (i.e. f is a linear map) which satisfies $\tau'(g)f = f\tau(g)$ for all $g \in \Gamma$. A $K\Gamma$-map which is bijective is a $K\Gamma$-*isomorphism*, or an *equivalence* between the representations τ, τ'. One has analogous definitions for right $K\Gamma$-modules; a right $K\Gamma$-module can be regarded as a pair (V, τ) where $\tau : \Gamma \to \mathsf{End}_K(V)$ is an *anti-representation* of Γ on the K-space V, i.e. $\tau(gg') = \tau(g')\tau(g)$ for all $g, g' \in \Gamma$, $\tau(1_\Gamma) = \mathbb{I}_V$.

The set K^Γ of all maps $\Gamma \to K$ is a commutative K-algebra, with algebra operations defined "pointwise", e.g. ff' is defined to take $g \mapsto f(g)f'(g)$, for every element ("point") g of Γ. The identity element $\mathbf{1}$ of K^Γ takes each $g \in \Gamma$ to the identity element 1_K of K. If $s \in \Gamma$ and $f \in K^\Gamma$, then the *left* and *right translates* of f by s are defined to be the maps $L_s f, R_s f : \Gamma \to K$ given by

$$L_s f : g \mapsto f(sg), \quad R_s f : g \mapsto f(gs), \quad g \in \Gamma.$$

Each of the operators L_s, R_s maps K^Γ into itself and is a K-algebra map (i.e. K-algebra homomorphism) $K^\Gamma \to K^\Gamma$. In particular, L_s, R_s both belong to the space $\mathsf{End}_K(K^\Gamma)$. It is easy to check that $R : s \mapsto R_s$ gives a representation of Γ on K^Γ, while $L : s \mapsto L_s$ gives an anti-representation. Thus K^Γ can be made into a left $K\Gamma$-module (using R) and a right $K\Gamma$-module (using L). We denote both module actions by $\circ$, so that if $s \in \Gamma$ and $f \in K^\Gamma$ we write

$$s \circ f = R_s f \quad \text{and} \quad f \circ s = L_s f.$$

Notice that these actions commute: $(s \circ f) \circ t = s \circ (f \circ t)$ for all $s, t \in \Gamma$ and $f \in K^\Gamma$. There is a linear map $K^\Gamma \otimes K^\Gamma \to K^{\Gamma\times\Gamma}$ ($\otimes$ means $\otimes_K$) which takes $f \otimes f'$ ($f, f' \in K^\Gamma$) to the function mapping $\Gamma \times \Gamma \to K$ by $(s,t) \mapsto f(s)f'(t)$, for all $s, t \in \Gamma$. This linear map is injective, *and we use it to identify* $K^\Gamma \otimes K^\Gamma$ *with a subspace of* $K^{\Gamma\times\Gamma}$.

The semigroup structure on Γ gives rise to two maps

$$\Delta : K^\Gamma \to K^{\Gamma\times\Gamma} \quad \text{and} \quad \varepsilon : K^\Gamma \to K,$$

as follows: if $f \in K^\Gamma$, then $\Delta f : (s,t) \mapsto f(st)$, and $\varepsilon(f) = f(1_\Gamma)$. Both Δ, ε are K-algebra maps. We shall say that an element $f \in K^\Gamma$ is *finitary*, or is a *representative function*, if it satisfies any one of the conditions F1, F2, F3 below: these three conditions are in fact equivalent (see e.g. [24, Chapter 2]).

F1. The left $K\Gamma$-submodule $K\Gamma \circ f$ generated by f is finite-dimensional.

F2. The right $K\Gamma$-submodule $f \circ K\Gamma$ generated by f is finite-dimensional.

F3. $\Delta f \in K^\Gamma \otimes K^\Gamma$. This means that there exist elements $f_h, f'_h \in K^\Gamma$ (where h runs over some *finite* index set) such that

(1a) $$\Delta f = \sum_h f_h \otimes f'_h.$$

This equation is equivalent to the system of equations

(1b) $$f(st) = \sum_h f_h(s) f'_h(t), \text{ all } s, t \in \Gamma.$$

It is also equivalent to each of the following systems

(1c) $$t \circ f = \sum_h f'_h(t) f_h, \text{ all } t \in \Gamma,$$

or

(1d) $$f \circ s = \sum_h f_h(s) f'_h, \text{ all } s \in \Gamma.$$

The set $F = F(K^\Gamma)$ of all finitary functions $f : \Gamma \to K$ is a K-bialgebra (see [51] for the definitions of coalgebras and bialgebras). It is a K-subalgebra of K^Γ, and is also closed to Δ in the sense that $\Delta F \subseteq F \otimes F$ (this means that if f is finitary, the functions f_h, f'_h in (1a) can be chosen to be themselves finitary). The K-space F, equipped with the maps $\Delta : F \to F \otimes F$, $\varepsilon : F \to K$, is a K-*coalgebra*; these two structures on F, of algebra and coalgebra, are linked by the fact that Δ and ε are both K-algebra maps (see [24, p. 15]).

Finitary functions on Γ appear as coefficient functions of finite-dimensional representations of Γ. Suppose τ is a representation of Γ on a finite-dimensional K-space V. If $\{ v_b : b \in B \}$ is a K-basis of V, we have equations

(1e) $$\tau(g) v_b = g v_b = \sum_{a \in B} r_{ab}(g) v_a, \text{ for } g \in \Gamma,\ b \in B;$$

here $r_{ab}(g) \in K$. The functions $r_{ab} : \Gamma \to K$ $(a, b \in B)$ are called *coefficient functions* of τ, or of the $K\Gamma$-module $V = (V, \tau)$. The K-span of these functions is a subspace of K^Γ called the *coefficient space*[1] of τ, or of the $K\Gamma$-module V. We denote this space by $\mathsf{cf}(V) = \sum_{a,b} K \cdot r_{ab}$; it is elementary to verify that it is independent of the choice of the basis $\{v_b\}$. The matrix $R = (r_{ab})$ gives a *matrix representation* of Γ, i.e. $R(gg') = R(g)R(g')$, $R(1_\Gamma) = (\delta_{ab})$ for all $g, g' \in \Gamma$ (δ_{ab} is the Kronecker delta). These conditions translate into conditions on the coefficients r_{ab}, viz.

(1f) $$\Delta r_{ab} = \sum_{c \in B} r_{ac} \otimes r_{cb}, \quad \varepsilon(r_{ab}) = \delta_{ab}, \text{ all } a, b \in B.$$

The matrix $R = (r_{ab})$ is sometimes called an "invariant matrix" [20, p. 140]. From the first equations it follows that all the coefficient functions r_{ab} are finitary, hence that $\mathsf{cf}(V)$ is a subspace of $F = F(K^\Gamma)$. But (1f) also shows that $C = \mathsf{cf}(V)$ is a *subcoalgebra* of F, i.e. that $\Delta C \subseteq C \otimes C$. As a matter of fact, every finitary function $f : \Gamma \to K$ lies in the coefficient space of some finite-dimensional $K\Gamma$-module V; for this purpose we could take $V = K\Gamma \circ f$ (see F1). It is for this reason that finitary functions are sometimes called "representative functions".

If S is any K-algebra (possibly of infinite dimension as K-space), $\mathsf{mod}(S)$ shall denote the category of all *finite-dimensional* left S-modules. Similarly, $\mathsf{mod}'(S)$ is the category of all finite-dimensional right S-modules. An *algebraic representation theory* of Γ over K could be defined as follows: first choose a subcoalgebra A of $F(K^\Gamma)$, i.e. A is a K-subspace of $F(K^\Gamma)$ satisfying $\Delta A \subseteq A \otimes A$. Then "$A$-representation theory" of Γ, is defined to be the study of the full subcategory $\mathsf{mod}_A(K\Gamma)$ of $\mathsf{mod}(K\Gamma)$, whose objects are all finite-dimensional left $K\Gamma$-modules V such that $\mathsf{cf}(V) \subseteq A$. (The morphisms $f : V \to V'$ between two objects V, V' of this category are, by definition, just the $K\Gamma$-maps.) In some contexts we say that a $K\Gamma$-module V is "rational", or more precisely "A-rational", if $\mathsf{cf}(V) \subseteq A$; then $\mathsf{mod}_A(K\Gamma)$ is the category of finite-dimensional A-rational left $K\Gamma$-modules. It is clear that submodules, quotient-modules and finite direct sums of A-rational modules, are themselves A-rational. We can define the category $\mathsf{mod}'_A(K\Gamma)$ of finite-dimensional right $K\Gamma$-modules which are A-rational in the same way. The assumption $\Delta A \subseteq A \otimes A$ implies that if $f \in A$, then the functions f_h, f'_h appearing in (1a) can themselves be chosen to belong to A. Then from (1c), (1d) follows that A is a left and right $K\Gamma$-submodule of K^Γ; also by quite elementary calculations that any finite-dimensional left (or right) $K\Gamma$-submodule V of A belongs to the category $\mathsf{mod}_A(K\Gamma)$ (or $\mathsf{mod}'_A(K\Gamma)$).

Examples.

1. Let Γ be an affine algebraic group over an algebraically closed field K (see for example [24, p. 21]), and $A = K[\Gamma]$ the ring of regular functions on Γ

[1] In [24], this is called the "space of representative functions" of τ, or V.

(A is often called the *affine ring* of Γ). Then $\mathsf{mod}_A(K\Gamma)$ is the category of rational (finite-dimensional) $K\Gamma$-modules in the usual sense of algebraic group theory. In this case, A is not only a subcoalgebra of $F(K^\Gamma)$, but a subbialgebra (see [49, p. 46]). The same remarks apply when Γ is an affine algebraic semigroup.

2. Let Γ be a finite semigroup, then of course $F(K^\Gamma) = K^\Gamma$. If we take $A = K^\Gamma$, then $\mathsf{mod}_A(K\Gamma) = \mathsf{mod}(K\Gamma)$. The (left and right) $K\Gamma$-module structures on A, are dual to the (right and left) "regular" $K\Gamma$-module structures on K given by multiplication: we may identify K^Γ with the dual space $(K\Gamma)^* = \mathsf{Hom}_K(K\Gamma, K)$.

3. Let K be an infinite field, n a positive integer, and $\Gamma = \mathsf{GL}_n(K)$, the group of all non-singular $n \times n$ matrices with coefficients in K. We could take $A = A_K(n)$, the ring of all polynomial functions $f : \Gamma \to K$ (see 2.1). The objects (V, τ) in $\mathsf{mod}_A(K\Gamma)$ (we shall later denote this category by $M_K(n)$, see 2.2) are called *polynomial $K\Gamma$-modules*, and the associated representations (including the matrix representations $R = (r_{ab})$ obtained by using the K-bases $\{v_b\}$ of V) are called *polynomial representations of* Γ. The study of such representations is the subject of these lectures. We get another category (denoted by $M_K(n, r)$ in 2.2) by taking $A = A_K(n, r)$, the space of polynomial functions on Γ which are homogeneous of degree r in the n^2 coefficients of a general element $g \in \Gamma$ (see 2.1 for a precise formulation). Finally we might mention that $A_K(n)$ can also be regarded as the affine ring of the algebraic semigroup $M_n(K)$ of *all* $n \times n$ matrices (singular or not) over K, so that we may regard polynomial representations of $\mathsf{GL}_n(K)$, as rational representations of $M_n(K)$, and conversely.

Now suppose once more that Γ is an arbitrary semigroup with identity 1_Γ, and that A is a subcoalgebra of the space $F(K^\Gamma)$ of all finitary functions on Γ. Then A is itself a coalgebra, relative to the maps $\Delta : A \to A \otimes A$ and $\varepsilon : A \to K$. So we may consider the category $\mathsf{com}(A)$ of all right A-comodules; an object V of $\mathsf{com}(A)$ is a finite-dimensional K-space, together with a "structure map" $\gamma : V \to V \otimes A$ which is K-linear and satisfies the identities $(\gamma \otimes \mathbb{I}_A)\gamma = (\mathbb{I}_V \otimes \Delta)\gamma$, $(\mathbb{I}_V \otimes \varepsilon)\gamma = \mathbb{I}_V$ (see [20, p. 138], where a right A-comodule is perversely referred to as a left A-comodule; better references are [24, p. 16], [49, p. 38] or [51, p. 30]). Our category $\mathsf{mod}_A(K\Gamma)$ is equivalent to $\mathsf{com}(A)$, as follows: if $V \in \mathsf{mod}_A(K\Gamma)$, take any K-basis $\{v_b\}$ of V and write down the equations (1e). Now define $\gamma : V \to V \otimes A$ to be the K-linear map given by equations

(1g) $$\gamma(v_b) = \sum_{a \in B} v_a \otimes r_{ab}, \text{ for } b \in B.$$

It is easy to check that γ is independent of the basis $\{v_b\}$. Moreover using (1f) we see that γ satisfies the comodule identities just given. Conversely given an A-comodule (V, γ), use equations (1g) to *define* the elements r_{ab} of A; the comodule identities now show that (1f) hold, so we may use (1e) to *define* the left $K\Gamma$-module $V = (V, \tau)$. It is evident that $\mathsf{cf}(V) \subseteq A$. So every A-rational, left $K\Gamma$-module can be regarded as a right A-comodule, and conversely. The definition of morphism $f : V \to V'$ in $\mathsf{com}(A)$ (see the references cited) is such that these morphisms are the same as $K\Gamma$-maps in $\mathsf{mod}_A(K\Gamma)$.

This formal transition from $K\Gamma$-modules to A-comodules is rather trivial, but it is nevertheless worth making, from several points of view. To begin with, the basic representation theory of arbitrary A-comodules (we should here work in the category $\mathsf{Com}(A)$, whose objects $V = (V, \gamma)$ are possibly infinite-dimensional) follows to a surprising extent the pattern discovered by R. Brauer and C. Nesbitt for finite-dimensional algebras (see [5, 20]). Included here is the possibility of a *modular theory*, which we shall discuss below.

Next, the A-comodule interpretation also permits us to profit by an important fact, namely that every right A-comodule can be regarded as a left module for the K-algebra $A^* = \mathsf{Hom}_K(A, K)$. The algebra structure in A^* is the dual of the coalgebra structure on A, i.e. if $\xi, \eta \in A^*$, we define the product[2] $\xi\eta$ to be the map of A into K which takes the element $f \in A$ to

(1h) $$\xi\eta(f) = \sum_h \xi(f_h)\eta(f'_h),$$

see (1a). The identity element of A^* is $\varepsilon : A \to K$. If $V = (V, \gamma)$ belongs to $\mathsf{com}(A)$, we make V into an A^*-module by the rule $\xi v = (\mathbb{I}_V \otimes \xi)(\gamma(v))$, for $\xi \in A^*$, $v \in V$. Working in terms of a basis $\{v_b\}$ of V, this rule becomes (see (1g))

(1i) $$\xi v_b = \sum_{a \in B} \xi(r_{ab})v_a, \text{ for } b \in B.$$

Therefore we have three kinds of matrix representation associated with our original $K\Gamma$-module $V = (V, \tau)$, relative to the basis $\{v_b\}$:

(i) the representation $g \mapsto (r_{ab}(g))$ of Γ;
(ii) the matrix $R = (r_{ab})$ whose elements are functions on Γ, satisfying equations (1f), and which can be thought of as a kind of representation of the coalgebra A;
(iii) the representation $\xi \mapsto (\xi(r_{ab}))$ of the algebra A^*, given by equations (1i).

We can recover (i) from (iii) very easily: for each $g \in \Gamma$ let $e_g : A \to K$ be "evaluation at g", i.e. $e_g(f) = f(g)$, for all $f \in A$. Then $e_g \in A^*$, and the map $e : \Gamma \to A^*$ satisfies $e_g e_{g'} = e_{gg'}$, $e_{1_\Gamma} = \varepsilon$, for $g, g' \in \Gamma$. So e may be extended linearly to a K-algebra map $e : K\Gamma \to A^*$, and if we compose the representation (iii) with e, we recover (i).

If A is finite-dimensional, then it is quite elementary to show that the two categories $\mathsf{mod}_A(K\Gamma)$ and $\mathsf{mod}(A^*)$ are equivalent; this amounts to showing that every finite-dimensional left A^*-module V yields a module in $\mathsf{mod}_A(K\Gamma)$ by composition with the map e. Schur exploited this fact in

[2]This product is often called "convolution".

the case $A = A_K(n,r)$, and could thereby work with the finite-dimensional algebra $A_K(n,r)^* = S_K(n,r)$ (which I have called the "Schur algebra" in these lectures, see 2.3, 2.4), instead of with the infinite-dimensional and irrelevantly complicated group algebra $K\Gamma$.

If A is infinite-dimensional, it is useful in many cases to regard modules $V \in \mathsf{mod}_A(K\Gamma)$ as modules over some "dense" subalgebra S of A^* (S is dense in A^* if, for every $0 \neq a \in A$, there is some $\xi \in S$ such that $\xi(a) \neq 0$). When $A = K[\Gamma]$ is the affine ring of a connected algebraic group Γ over an algebraically closed field K, one may take S the "hyperalgebra" $\mathsf{hy}(\Gamma)$ of Γ (see [9, §6]). In case Γ is simply-connected and semisimple, the correspondence between $\mathsf{mod}_A(K\Gamma)$ and $\mathsf{mod}(S)$ sets up an equivalence of categories (J. Sullivan; see [9, 6.8]). Moreover in that case $\mathsf{hy}(\Gamma)$ can be identified with an algebra U_K constructed out of the complex semisimple Lie algebra associated with the root system of Γ (W. Haboush; [9, 6.5, 6.6] or [22, 1.3]). This algebra U_K (which is sometimes *defined* to be the hyperalgebra of Γ) has an explicit basis with sufficiently good multiplicative properties to make it immensely valuable in studying the rational representations of Γ. In an important paper [6] R. Carter and G. Lusztig have used the hyperalgebra—rather than the Schur algebra—to investigate the polynomial representations of $\Gamma = \mathsf{GL}_n(K)$.

Carter-Lusztig use the idea, which is derived from C. Chevalley's fundamental paper [7] on split semisimple algebraic groups, that the family of all groups $\mathsf{GL}_n(K)$ (n fixed, K varying over some class $\mathcal{K}$ of commutative rings) is "defined over $\mathbb{Z}$". This makes possible a "modular theory" for the polynomial representations of these groups, which in its essentials corresponds to R. Brauer's modular representation theory for finite groups. We can give a sufficiently general setting for such a theory as follows. Suppose we have a family $\{\Gamma_K, A_K\}$, where for each K in the class $\mathcal{K}$ of all infinite fields, Γ_K is a semigroup and A_K is a K-subcoalgebra of $F(K^{\Gamma_K})$. Suppose also that the following two conditions are satisfied. ($\mathbb{Q}$ denotes the rational field.)

Z1. The $\mathbb{Q}$-coalgebra $A_\mathbb{Q} = (A_\mathbb{Q}, \Delta_\mathbb{Q}, \varepsilon_\mathbb{Q})$ contains a $\mathbb{Z}$*-form* $A_\mathbb{Z}$, i.e. (a) $A_\mathbb{Z}$ is a lattice in $A_\mathbb{Q}$, which means $A_\mathbb{Z} = \sum_\nu \mathbb{Z}a_\nu$ for some $\mathbb{Q}$-basis $\{a_\nu\}$ of $A_\mathbb{Q}$, and (b) $\Delta_\mathbb{Q}(A_\mathbb{Z}) \subseteq A_\mathbb{Z} \otimes A_\mathbb{Z}$, $\varepsilon_\mathbb{Q}(A_\mathbb{Z}) \subseteq \mathbb{Z}$.

Z2. For each $K \in \mathcal{K}$ there is a K-coalgebra isomorphism $\alpha_K : A_\mathbb{Z} \otimes K \to A_K$ (here $\otimes$ means $\otimes_\mathbb{Z}$, and $A_\mathbb{Z} \otimes K$ is made into a K-coalgebra by "extension of scalars").

In this case we say that the family $\{\Gamma_K, A_K\}$ is *defined over* $\mathbb{Z}$ by means of $A_\mathbb{Z}$.

Examples.

4. Let $\pi : \mathcal{G}_\mathbb{C} \to \mathsf{End}_\mathbb{C} E$ be a faithful representation of a complex semisimple Lie algebra $\mathcal{G}_\mathbb{C}$ over a complex vector space E of finite dimension n, and let $E_\mathbb{Z}$ be an "admissible lattice" in E (see [4, p. A-5] or [50, p. 17]). For each $K \in \mathcal{K}$ let Γ_K be the Chevalley group over K defined by π, $E_{\mathbb{Z}}$; its elements can be regarded as matrices $g = (g_{\mu\nu})$ in $\mathsf{SL}_n(K)$. For each pair (μ, ν) define the coefficient function $c^K_{\mu\nu} : g \mapsto g_{\mu\nu}$. From the equations $\Delta c^K_{\mu\nu} = \sum_\lambda c^K_{\mu\lambda} \otimes c^K_{\lambda\nu}$, we deduce that the K-subalgebra generated by all the $c^K_{\mu\nu}$ is a K-subcoalgebra (hence even a K-subbialgebra) of $F(K^{\Gamma_K})$; we take this to be A_K. Chevalley showed in [7] (see also [4, §4]) that the family $\{\Gamma_K, A_K\}$ is defined over $\mathbb{Z}$. The relevant $\mathbb{Z}$-form $A_{\mathbb{Z}}$ of $A_{\mathbb{Q}}$ is just the subring of $A_{\mathbb{Q}}$ generated by the $c^{\mathbb{Q}}_{\mu\nu}$; the maps α_K are K-algebra (as well as K-coalgebra) isomorphisms, and take $c^{\mathbb{Q}}_{\mu\nu} \otimes 1_K \mapsto c^K_{\mu\nu}$ for all μ, ν. (From the standpoint of algebraic group theory, each pair (Γ_K, A_K) is an affine algebraic group defined over K, and the family $\{\Gamma_K, A_K\}$ is an "affine group scheme over $\mathbb{Z}$", defined by the $\mathbb{Z}$-bialgebra $A_{\mathbb{Z}}$. See [49, p. 46].)

5. Fix a positive integer n, and let $\Gamma_K = \mathsf{GL}_n(K)$ for each $K \in \mathcal{K}$. For A_K we may take either $A_K(n)$, or $A_K(n, r)$ for some fixed $r \geq 0$ (see 2.1). It is completely elementary to verify that in each case the family $\{\Gamma_K, A_K\}$ is defined over $\mathbb{Z}$; the relevant $\mathbb{Z}$-forms $A_{\mathbb{Z}}(n)$, $A_{\mathbb{Z}}(n, r)$ are described in 2.5. In these lectures, we study the family $\{\Gamma_K, A_K(n, r)\}$.

The first essential of the *modular representation theory* of any family $\{\Gamma_K, A_K\}$ which is defined over $\mathbb{Z}$, is the process of *modular reduction.* We shall write M_K for the category $\mathsf{mod}_{A_K}(K\Gamma_K)$, for any $K \in \mathcal{K}$. Then an object $V_{\mathbb{Q}}$ in $M_{\mathbb{Q}}$ is a finite-dimensional $\mathbb{Q}$-space on which $\Gamma_{\mathbb{Q}}$ acts. If $\{ v_{b,\mathbb{Q}} : b \in B \}$ is a $\mathbb{Q}$-basis of $V_{\mathbb{Q}}$, we have equations like (1e)

$$\text{(1j)} \quad g v_{b,\mathbb{Q}} = \sum_{a \in B} r^{\mathbb{Q}}_{ab}(g) v_{a,\mathbb{Q}}, \text{ for } g \in \Gamma_{\mathbb{Q}},\ b \in B.$$

Here the functions $r^{\mathbb{Q}}_{ab}$ belong to $A_{\mathbb{Q}}$, and satisfy equations like (1f). We make the following definition: a subset $V_{\mathbb{Z}}$ of $V_{\mathbb{Q}}$ is called a *$\mathbb{Z}$-form* (or *admissible lattice*) of $V_{\mathbb{Q}}$ if

(a) $V_{\mathbb{Z}}$ is a lattice in $V_{\mathbb{Q}}$, which means $V_{\mathbb{Z}} = \sum_b \mathbb{Z} v_{b,\mathbb{Q}}$ for some $\mathbb{Q}$-basis $\{v_{b,\mathbb{Q}}\}$ of $V_{\mathbb{Q}}$, and

(b) All the coefficient functions $r^{\mathbb{Q}}_{ab}$, relative to this basis, lie in $A_{\mathbb{Z}}$.

Another way of expressing condition (b) is to convert $V_{\mathbb{Q}}$ into an $A_{\mathbb{Q}}$-comodule by means of the map $\gamma_{\mathbb{Q}} : V_{\mathbb{Q}} \to V_{\mathbb{Q}} \otimes A_{\mathbb{Q}}$, using equations like (1g). Then (b) is equivalent to

(b') $\gamma_{\mathbb{Q}}(V_{\mathbb{Z}}) \subseteq V_{\mathbb{Z}} \otimes A_{\mathbb{Z}}$.

Now suppose that $K \in \mathcal{K}$. We can make the K-space $V_K = V_{\mathbb{Z}} \otimes K$ (here $\otimes$ means $\otimes_{\mathbb{Z}}$) into an object of M_K, as follows. Define $r^K_{ab} = \alpha_K(r^{\mathbb{Q}}_{ab} \otimes 1_K) \in A_K$, using the K-coalgebra isomorphism $\alpha_K : A_{\mathbb{Z}} \otimes K \to A_K$ postulated in Z2. These r^K_{ab} satisfy equations like (1f). So we may define an action of Γ_K on V_K by equations

(1k) $$gv_{b,K} = \sum_{a\in B} r_{ab}^{K}(g) v_{a,K}, \text{ for } g \in \Gamma_K, b \in B.$$

Here $v_{b,K} = v_{b,\mathbb{Q}} \otimes 1_K$, for $b \in B$. The process by which $V_{\mathbb{Q}}$ is converted, via the $\mathbb{Z}$-form $V_{\mathbb{Z}}$, into V_K is called modular reduction. A general theorem guarantees that each $V_{\mathbb{Q}} \in M_{\mathbb{Q}}$ possesses at least one $\mathbb{Z}$-form $V_{\mathbb{Z}}$ (see [49, Lemma 2, p. 43] or [20, (2.2d), p. 159]). Different $\mathbb{Z}$-forms $V_{\mathbb{Z}}, V'_{\mathbb{Z}}, \ldots$ of the same $V_{\mathbb{Q}}$ may give non-isomorphic $V_K = V_{\mathbb{Z}} \otimes K, V'_K = V'_{\mathbb{Z}} \otimes K, \ldots$ in M_K, but another general theorem (due in its original form to Brauer and Nesbitt) says that all these modules $V_K, V'_K, \ldots$ have the same composition factor multiplicities; from this the notion of *decomposition numbers* can be defined (see [49, p. 44] or [20, (2.5a), p. 162]).

In these lectures we take $\Gamma = \mathsf{GL}_n(K)$, where K is an infinite field, and study $K\Gamma$-modules $V = (V, \tau)$, which belong to the category $M_K(n,r)$, for a fixed homogeneity degree r (see Example 3, above). In chapter 2 the Schur algebra $S_K(n,r)$ is defined, and it is shown how $K\Gamma$-modules in $M_K(n,r)$ can be regarded as left $S_K(n,r)$-modules, and conversely. An alternative description of $S_K(n,r)$ is that it is the endomorphism algebra of the rth tensor space $E^{\otimes r}$, when the latter is given its natural structure as a module for the symmetric group $G(r)$. This has as corollary Schur's theorem (2.6e): if $\mathsf{char}\, K = 0$, then every module V in $M_K(n,r)$ is completely reducible.

Schur's multiplication rule for $S_K(n,r)$ (see (2.3b)) provides an effective method for calculating with modules in $M_K(n,r)$. For example, the "weight spaces" of such a module V are easily expressed in terms of certain idempotent elements ξ_α in $S_K(n,r)$. Weights and characters are discussed in chapter 3. By definition, the character of V is a symmetric polynomial over $\mathbb{Z}$, which is homogeneous of degree r in a set of n variables $X_1, \ldots, X_n$. In 3.5 is reproduced the argument by which Schur showed that the isomorphism classes of irreducible modules in $M_K(n,r)$ are in one-one correspondence with the partitions $\lambda = (\lambda_1, \ldots, \lambda_n)$ of r into not more than n parts. Of course Schur considered only the case $K = \mathbb{C}$, but his argument requires only minor modification for an arbitrary infinite field K. The character of an irreducible module of type λ depends only on the characteristic p of K; we write this $\phi_{\lambda,p}$. For $p \neq 0$ these characters have not yet been determined except in special cases. For $p = 0$, Schur showed in [47] that they are the symmetric functions now known as "Schur functions". A proof of this is given at then end of 3.5—our proof uses some identities involving symmetric functions which can be found, for example, in I. G. Macdonald's recent book [39].

In chapters 4 and 5, I have departed widely from Schur's dissertation. These chapters are concerned with the construction, for each λ and for each K, of two modules $D_{\lambda,K}$ and $V_{\lambda,K}$ in our category $M_K(n,r)$. They are "explicit" in the sense that a basis can be given for each. They are dual to each other, in the sense of the "contravariant" duality described in 2.7. $V_{\lambda,K}$ has a unique irreducible factor module; this is denoted $F_{\lambda,K}$. $D_{\lambda,K}$ has a unique minimal submodule, which is isomorphic to $F_{\lambda,K}$. The set $\{F_{\lambda,K}\}$, as λ ranges over all partitions λ of r into not more than n parts, gives a full set of irreducible modules in $M_K(n,r)$. If $\mathsf{char}\, K = 0$, then $F_{\lambda,K} \cong V_{\lambda,K} \cong D_{\lambda,K}$. But for $\mathsf{char}\, K = p \neq 0$, knowledge of $F_{\lambda,K}$ is still very incomplete.

The history of the modules $D_{\lambda,K}$ and $V_{\lambda,K}$ is interesting, and I am indebted to J. Towber (see [52]) for much of the following information. $D_{\lambda,K}$ is generated by certain determinantal expressions (here denoted $(T_l : T_i)$), whose significance as "primary covariants" was noted by J. Deruyts [13], in 1892. Although Schur refers to two later papers of Deruyts, there is no sign in [47] that he appreciated that Deruyts had really given a complete set of irreducible modules in $M_{\mathbb{C}}(n,r)$. The discovery of the basis of the "standard" $(T_l : T_i)$, seems to go back to A. Young [58, 1902]. The observation that the $D_{\lambda,K}$ can be constructed over an arbitrary field—or equivalently that the $(T_l : T_i)$ generate a $\mathbb{Z}$-form $D_{\lambda,\mathbb{Z}}$ in $D_{\lambda,\mathbb{Q}}$—was made by G. Higman [23, 1965]. The $V_{\lambda,K}$ (and the $\mathbb{Z}$-form $V_{\lambda,\mathbb{Z}}$) were constructed, independently of all this, by R. Carter and G. Lusztig [6, 1974]. They called these "Weyl modules", and their construction was based on methods used in the theory of semisimple algebraic groups. Towber [52] showed that $D_{\lambda,K}$ and $V_{\lambda,K}$ are dual to each other—his framework is "functorial" and more general than ours. M. Clausen [8] has used recent combinatorial theory of G.-C. Rota and his collaborators [15] to construct both modules $D_{\lambda,K}$ and $V_{\lambda,K}$. G. D. James describes, in his book [27], some $K\Gamma$-modules which are isomorphic to the $D_{\lambda,K}$. His construction is quite different from those above; we show in 4.8 that it yields the important and deep fact that $D_{\lambda,K}$ is an "induced" module, in the sense of algebraic group theory.

Chapter 6 returns to Schur's dissertation. I have "reversed" the elegant procedure by which he constructed $K\Gamma$-modules from modules for the symmetric group $G(r)$. This provides an interesting illumination of some recent work of James on the modular representation theory of the symmetric group.

2

Polynomial Representations of $\mathsf{GL}_n(K)$: The Schur algebra

2.1 Notation, etc.

Let n be a positive integer, K an infinite field, and $\Gamma = \mathsf{GL}_n(K)$ the group of all non-singular $n \times n$ matrices over K. For each pair μ, ν of elements of $\underline{n} = \{1, \ldots, n\}$, let $c_{\mu\nu} \in K^\Gamma$ be the function which associates to each $g \in \Gamma$ its (μ, ν)-coefficient $g_{\mu\nu}$. Denote by A or $A_K(n)$ the K-subalgebra of K^Γ generated by the functions $c_{\mu\nu}$ $(\mu, \nu \in \underline{n})$; the elements of A are, by definition, the polynomial functions on Γ. Since K is infinite, the $c_{\mu\nu}$ are algebraically independent over K, so that A can be regarded as the algebra of all polynomials over K in n^2 "indeterminates" $c_{\mu\nu}$ $(\mu, \nu \in \underline{n})$.

For each $r \geq 0$ we denote by $A_K(n,r)$ the subspace of A consisting of the elements expressible as polynomials which are homogeneous of degree r in the $c_{\mu\nu}$. Then $A_K(n,r)$ has finite dimension $\binom{n^2+r-1}{r}$ as K-space; in particular $A_K(n,0) = K \cdot 1_A$, where 1_A denotes the constant function 1_A which maps $g \mapsto 1_K$ for all $g \in \Gamma$. The K-algebra A has the standard grading

(2.1a) $$A = A_K(n) = \bigoplus_{r \geq 0} A_K(n,r).$$

If integers n, r (both ≥ 1) are given, we write $I(n,r)$ for the set of all functions $i : \underline{r} \to \underline{n}$. Such a function is usually written as vector or "multi-index" $i = (i_1, \ldots, i_r)$ with values $i_\varrho \in \underline{n}$. The symmetric group on the set $\underline{r} = \{1, \ldots, r\}$ is denoted $G(r)$ or G. It acts naturally on the *right* on $I(n,r)$ by $i\pi = (i_{\pi(1)}, \ldots, i_{\pi(r)})$, so that $\pi \in G(r)$ acts as "place-permutation" on each $i \in I(n,r)$. We make $G(r)$ act also on the set $I(n,r) \times I(n,r)$ by $(i,j)\pi = (i\pi, j\pi)$. We write $i \sim j$ to indicate that the elements i, j of $I(n,r)$ are in the same $G(r)$-orbit, i.e. that $j = i\pi$ for some $\pi \in G(r)$. Similarly $(i,j) \sim (k,l)$ means that $k = i\pi$ and $l = j\pi$ for some $\pi \in G(r)$.

As an example of the use of this notation, notice that $A_K(n,r)$ is spanned, as K-space, by the monomials

(2.1b) $$c_{i,j} = c_{i_1 j_1} c_{i_2 j_2} \cdots c_{i_r j_r},$$

for all $i, j \in I(n,r)$. Of course the pair (i,j) is not uniquely determined by the monomial (2.1b); in fact $c_{i,j} = c_{k,l}$ if and only if $(i,j) \sim (k,l)$. The space $A_K(n,r)$ has as K-basis the set of distinct monomials (2.1b), and these are in bijective correspondence with the $G(r)$-orbits of $I(n,r) \times I(n,r)$. Thus the number of these orbits is $\binom{n^2+r-1}{r}$.

2.2 The categories $\mathbf{M_K(n)}$, $\mathbf{M_K(n,r)}$

The maps $\Delta : K^\Gamma \to K^{\Gamma\times\Gamma}$, $\varepsilon : K^\Gamma \to K$ (see introduction) behave as follows on the functions $c_{\mu\nu}$ $(\mu, \nu \in \underline{n})$:

(2.2a) $$\Delta(c_{\mu\nu}) = \sum_{\lambda\in\underline{n}} c_{\mu\lambda} \otimes c_{\lambda\nu}, \quad \varepsilon(c_{\mu\nu}) = \delta_{\mu\nu}.$$

These follow from the rule for multiplying two matrices, and from the formula for the unit element 1_Γ of Γ. Since Δ, ε are both multiplicative we deduce, for any "multi-indices" $p, q \in I(n,r)$ of length $r \geq 1$,

(2.2b) $$\Delta(c_{p,q}) = \sum_{s\in I} c_{p,s} \otimes c_{s,q}, \quad \varepsilon(c_{p,q}) = \delta_{p,q}.$$

Here $\delta_{p,q} = 1$ or 0, according as $p = q$ or $p \neq q$. These formulae show that $A = A_K(n)$ is a subcoalgebra (hence also a subbialgebra) of $F(K^\Gamma)$, and that each $A_K(n,r)$ is a subcoalgebra of $A_K(n)$ (for $r = 0$ this is because $\Delta 1_A = 1_A \otimes 1_A$). We shall write $M_K(n)$ and $M_K(n,r)$ for the categories $\mathsf{mod}_{A_K(n)}(K\Gamma)$ and $\mathsf{mod}_{A_K(n,r)}(K\Gamma)$. Thus $M_K(n)$ is the category of finite-dimensional (left) $K\Gamma$-modules which afford "polynomial" representations of $\Gamma = \mathsf{GL}_n(K)$; and $M_K(n,r)$ is the subcategory consisting of those affording representations in which all the coefficients are polynomials homogeneous of degree r in the $c_{\mu\nu}$. By an argument first given by Schur [47, p. 5] in case $K = \mathbb{C}$, but valid for any infinite field K of any characteristic, we have

(2.2c) Theorem. *Each $K\Gamma$-module $V \in M_K(n)$ has a direct sum decomposition*

$$V = \bigoplus_{r\geq 0} V_r\,,$$

where for each $r \geq 0$, V_r is a sub-module of V with $\mathsf{cf}(V_r) \leq A_K(n,r)$, that is $V_r \in M_K(n,r)$.

In other words each polynomial representation of Γ is equivalent to a direct sum of homogeneous ones.

Remark. In fact (2.2c) follows from a general theorem on A-comodules, where A is any coalgebra which is a direct sum $A = \bigoplus_\varrho A_\varrho$ of subcoalgebras A_ϱ (ϱ ranging over an index set P). This theorem says that if $V = (V, \tau) \in \mathsf{com}(A)$, then $V = \bigoplus_\varrho V_\varrho$, where for each $\varrho \in P$ the space V_ϱ is the unique maximum sub-comodule of V such that $\mathsf{cf}(V_\varrho) \leq A_\varrho$, i.e. such that $V_\varrho \in \mathsf{com}(A_\varrho)$ [20, p. 156, (1.6c). The proof given there does not depend on the assumption that the subcoalgebra summands R_ϱ are minimal].

2.3 The Schur algebra $\mathbf{S_K(n,r)}$

Theorem (2.2c) shows that each *indecomposable* module $V \in M_K(n)$ is homogeneous, i.e. $V \in M_K(n,r)$ for some $r \geq 0$. This means that we may as well confine our attention to homogeneous modules. From now on, let $r \geq 0$ be fixed, and define $S_K(n,r)$ to be the dual space of $A_K(n,r)$:

$$S_K(n,r) = A_K(n,r)^* = \mathsf{Hom}_K(A_K(n,r), K).$$

As K-space, $S_K(n,r)$ has basis $\{\xi_{i,j} : i,j \in I(n,r)\}$ dual to the basis $\{c_{i,j} : i,j \in I(n,r)\}$ of $A_K(n,r)$. For $i,j \in I(n,r)$, $\xi_{i,j}$ is the element of $S_K(n,r)$ given by

$$\xi_{i,j}(c_{p,q}) = \begin{cases} 1 & \text{if } (i,j) \sim (p,q) \\ 0 & \text{if } (i,j) \not\sim (p,q) \end{cases}, \quad \text{all } p,q \in I(n,r).$$

As with the $c_{i,j}$ we have an *equality rule* to take into account: $\xi_{i,j} = \xi_{k,l}$ if and only if $(i,j) \sim (k,l)$. The dimension of $S_K(n,r)$ is of course equal to $\binom{n^2+r-1}{r} = \dim A_K(n,r)$.

Since $A_K(n,r)$ is a coalgebra, its dual $S_K(n,r)$ is an associative algebra. We saw in §1 that the product $\xi\eta$ of elements ξ, η of $S_K(n,r)$ is defined as follows: if $c \in A_K(n,r)$ and if

$$\Delta(c) = \sum_t c_t \otimes c'_t$$

where the sum is finite and the $c_t, c'_t \in A_K(n,r)$, then

(2.3a) $$(\xi\eta)(c) = \sum_t \xi(c_t)\eta(c'_t).$$

The unit element of $S_K(n,r)$ will be denoted ε; it is given by $\varepsilon(c) = c(1_\Gamma)$ for all $c \in A_K(n,r)$.

Applying (2.3a) to a basis element $c = c_{p,q}$ of $A_K(n,r)$, we get (see (2.2b))

$$(\xi\eta)(c_{p,q}) = \sum_{s \in I(n,r)} \xi(c_{p,s})\eta(c_{s,q}).$$

Specializing to the case where $\xi = \xi_{i,j}$, $\eta = \xi_{k,l}$ are basis elements of $S_K(n,r)$, we deduce a

Multiplication Rule for $\mathbf{S_K(n,r)}$.

(2.3b) $$\xi_{i,j}\xi_{k,l} = \sum_{p,q} \{Z(i,j,k,l,p,q).1_K\}\xi_{p,q},$$

where the sum is over a set of representatives (p,q) of the $G(r)$-orbits of $I(n,r) \times I(n,r)$, and

$$Z(i,j,k,l,p,q) = \mathsf{Card}\{\, s \in I(n,r) \,:\, (i,j) \sim (p,s) \text{ and } (k,l) \sim (s,q) \,\}.$$

This multiplication rule (rather differently expressed) is due to Schur (see [47, p. 20]). Some special cases are worth noticing.

(2.3c) *For any $i, j, k, l \in I(n,r)$ there hold*

(i) $\xi_{i,j}\xi_{k,l} = 0$ *unless* $j \sim k$, *and*

(ii) $\xi_{i,i}\xi_{i,j} = \xi_{i,j} = \xi_{i,j}\xi_{j,j}$.

For example, (i) holds because if $\xi_{i,j}\xi_{k,l} \neq 0$, then by (2.3b) there must exist s, p, q with $(i,j) \sim (p,s)$ and $(k,l) \sim (s,q)$. This implies $j \sim s$ and $k \sim s$, hence $j \sim k$.

From (2.3c) follows that $\xi_{i,i}^2 = \xi_{i,i}$, and $\xi_{i,i}\xi_{j,j} = 0$ if $i \not\sim j$. Of course if $i \sim j$, then $(i,i) \sim (j,j)$ and hence $\xi_{i,i} = \xi_{j,j}$. But the distinct $\xi_{i,i}$ form a set of mutually orthogonal idempotents, and their sum is the unit element ε of $S_K(n,r)$. This last equation,

(2.3d) $\varepsilon = \sum_i \xi_{i,i}$, sum over a set of representatives of the $G(r)$-orbits of $I(n,r)$

is proved by evaluating both sides at all the basis elements $c_{p,q}$ of $A_K(n,r)$.

Of great importance in the modular theory for GL_n is the fact that, for fixed n, r, the scheme or family of algebras $S_K(n,r)$ is "defined over $\mathbb{Z}$", in the following sense. Let us use a superscript K to denote the basis elements $\xi_{i,j}^K$ of $S_K(n,r)$. It is clear from (2.3b) that the $\mathbb{Z}$-submodule $S_\mathbb{Z}(n,r)$ of $S_\mathbb{Q}(n,r)$, which is generated by the $\xi_{i,j}^\mathbb{Q}$ $(i,j \in I(n,r))$, is multiplicatively closed—it is a $\mathbb{Z}$-*order* in $S_\mathbb{Q}(n,r)$. And for any field K, there is an isomorphism of K-algebras $S_\mathbb{Z}(n,r) \otimes K \cong S_K(n,r)$ which takes each $\xi_{i,j}^\mathbb{Q} \otimes 1_K \mapsto \xi_{i,j}^K$.

2.4 The map $\mathbf{e} : \mathbf{K\Gamma} \to \mathbf{S_K(n,r)}$

For each $g \in \Gamma$ we define the element $e_g \in S_K(n,r)$ by $e_g(c) = c(g)$ for all $c \in A_K(n,r)$. It is clear from (2.3a) and §1 that $e_g e_{g'} = e_{gg'}$ for all $g, g' \in \Gamma$; also $e_1 = \varepsilon$ by the definition of ε. So if we extend the map $g \mapsto e_g$ linearly we get a map $e : K\Gamma \to S_K(n,r)$ which is a morphism of K-algebras.

Any function $f \in K^\Gamma$ has a unique extension to a linear map $f : K\Gamma \to K$. With this convention, the image under e of an element $\kappa = \sum \kappa_g g \in K\Gamma$, is "evaluation at κ"; i.e.

(2.4a) $e(\kappa) : c \mapsto c(\kappa)$, all $c \in A_K(n,r)$.

Propositions (2.4b), (2.4c) give the most important facts about e.

(2.4b) Proposition.

(i) e *is surjective.*

(ii) *Let $Y = \mathsf{Ker}\, e$, and let f be any element of K^Γ. Then $f \in A_K(n,r)$ if and only if $f(Y) = 0$.*

Proof. (i) If $\mathsf{Im}\, e$ were a proper subspace of $S_K(n,r) = A_K(n,r)^*$, there would exist some $0 \neq c \in A_K(n,r)$ such that $e_g(c) = c(g) = 0$ for all $g \in \Gamma$, a contradiction.

(ii) If $f \in A_K(n,r)$ and $\kappa \in Y$, we have $e(\kappa) = 0$ and hence $f(\kappa) = 0$, by (2.4a). So $f(Y) = 0$. Now suppose conversely that f is any element of K^Γ such that $f(Y) = 0$. By (i) there is an exact sequence

$$0 \longrightarrow Y \longrightarrow K\Gamma \xrightarrow{e} S_K(n,r) \longrightarrow 0,$$

from which it is clear that there exists an element $y \in S_K(n,r)^*$ such that $y(e(\kappa)) = f(\kappa)$ for all $\kappa \in K\Gamma$. By the natural isomorphism

$$S_K(n,r)^* \cong A_K(n,r),$$

there exists $c \in A_K(n,r)$ such that $y(\xi) = \xi(c)$ for all $\xi \in S_K(n,r)$. Put $\xi = e(\kappa)$, then we have $f(\kappa) = e(\kappa)(c) = c(\kappa)$, all $\kappa \in K\Gamma$. Therefore $f = c$, and the proof of (2.4b) is complete.

(2.4c) Proposition. *Let $V \in \mathsf{mod}(K\Gamma)$. Then $V \in M_K(n,r)$ if and only if $YV = 0$.*

Proof. Let $\{v_b\}$ be a basis of V, and (r_{ab}) the invariant matrix afforded by the action of $K\Gamma$ on this basis (see §1). Clearly $YV = 0$ if and only if $r_{ab}(Y) = 0$ for all a, b. By the last proposition, this is equivalent to saying that all the r_{ab} lie in $A_K(n,r)$, that is, that $\mathsf{cf}(V) \leq A_K(n,r)$. But of course this is the condition for V to belong to $M_K(n,r)$, and so the proof of (2.4c) is complete.

These propositions show that *the categories $M_K(n,r)$ and $\mathsf{mod}(S_K(n,r))$ are equivalent*, and in a very elementary way; an object V in either category can be transformed into an object of the other, using the rule

(2.4d) $\quad \kappa v = e(\kappa)v$, all $\kappa \in K\Gamma$, $v \in V$

to relate the action on V of the two algebras $K\Gamma$ and $S_K(n,r)$. Since both actions determine the same algebra of linear transformations on V, the concepts of submodule, module homomorphism, etc. coincide in the two categories. This category equivalence was one of the main techniques used by Schur in his dissertation [47, p. 21]. We might mention that the action of $S_K(n,r) = A_K(n,r)^*$ on a module $V \in M_K(n,r)$, which is given by (2.4d), is the same as that which is obtained by the general procedure outlined in the introduction. If the action of Γ on a basis $\{v_b\}$ of V is given by equations (1e), then the action of $S_K(n,r)$ is given by

$$\xi v_b = \sum_a \xi(r_{ab}) v_a\,, \text{ all } \xi \in S_K(n,r),\ b \in B.$$

For it is clear that (2.4d) holds, whenever $\kappa = g$ and $v = v_b$. By linearity it holds for all $\kappa \in K\Gamma$ and $v \in V$.

2.5 Modular theory

R. Brauer's theory of modular representations of finite groups was extended, by Brauer himself [5] and by Nakayama [42, 43, 44], to finite dimensional algebras. More recently Serre [49] (see also [20]) extended the theory to affine algebraic groups, or rather to affine algebraic group schemes. (The group scheme GL_n is a functor, which associates to each commutative ring K the group $\Gamma_K = \mathsf{GL}_n(K)$.) We can describe the characteristic modular "reduction" or "decomposition" process as follows.

Let $A_\mathbb{Z}(n)$, $A_\mathbb{Z}(n,r)$ be the subsets of $A_\mathbb{Q}(n)$, $A_\mathbb{Q}(n,r)$ respectively, consisting of those polynomials in the $c_{\mu\nu}$ whose coefficients all lie in $\mathbb{Z}$. These are "$\mathbb{Z}$-forms" of $A_\mathbb{Q}(n)$, $A_\mathbb{Q}(n,r)$; for example, $A_\mathbb{Z}(n,r)$ is the $\mathbb{Z}$-span of the $\mathbb{Q}$-basis $\{c^\mathbb{Q}_{i,j}\}$ of $A_\mathbb{Q}(n,r)$, and we have $\Delta A_\mathbb{Z}(n,r) \subseteq A_\mathbb{Z}(n,r) \otimes A_\mathbb{Z}(n,r)$ and $\varepsilon(A_\mathbb{Z}(n,r)) \leq \mathbb{Z}$ (see (2.2b)). For any infinite field K, there is a K-coalgebra isomorphism $A_\mathbb{Z}(n,r) \otimes K \cong A_K(n,r)$ which takes $c^\mathbb{Q}_{i,j} \otimes 1_K \mapsto c^K_{i,j}$ for all $i,j \in I(n,r)$. The $\mathbb{Z}$-order $S_\mathbb{Z}(n,r)$ which we defined in 2.3, is the set of all elements $\xi \in S_\mathbb{Q}(n,r)$ such that $\xi(A_\mathbb{Z}(n,r)) \leq \mathbb{Z}$.

Now let $V_\mathbb{Q}$ be any object in $M_\mathbb{Q}(n,r)$; we shall regard $V_\mathbb{Q}$ as a module for $S_\mathbb{Q}(n,r)$ when this is convenient. By a *$\mathbb{Z}$-form* of $V_\mathbb{Q}$ is meant a subset $V_\mathbb{Z}$ which

(i) is the $\mathbb{Z}$-span of some $\mathbb{Q}$-basis $\{v_b\}$ of $V_\mathbb{Q}$, and
(ii) is closed to the action of $S_\mathbb{Z}(n,r)$.

If $R_\mathbb{Q} = (r_{ab})$ is the invariant matrix defined by $\{v_b\}$ (see (1e)), then condition (ii) just says that all the r_{ab} lie in $A_\mathbb{Z}(n,r)$. Still another formulation of (ii) is that $\tau(V_\mathbb{Z}) \subseteq V_\mathbb{Z} \otimes A_\mathbb{Z}(n,r)$, where $(V_\mathbb{Q}, \tau)$ is the $A_\mathbb{Z}(n,r)$-comodule determined by $V_\mathbb{Q}$. That every $\mathbb{Q}\Gamma_\mathbb{Q}$-module $V_\mathbb{Q}$ in $M_\mathbb{Q}(n,r)$ contains at least one $\mathbb{Z}$-form, follows from [5, p. 256, §6] or [49, p. 43, lemme 2] or [20, p. 158, (2.2c)].

Now take any infinite field K. It is clear that the K-space $V_K = V_\mathbb{Z} \otimes K$ can be regarded as a left module for $S_K(n,r) \cong S_\mathbb{Z}(n,r) \otimes K$, hence as a $K\Gamma_K$-module in $M_K(n,r)$. The transition from $V_\mathbb{Q}$ to V_K is particularly easy to express in terms of invariant matrices; the invariant matrix R_K defined by the K-basis $\{v_b \otimes 1_K\}$ of V_K, is $(r_{ab} \otimes 1_K)$, where $(r_{ab}) = R_\mathbb{Q}$ is the invariant matrix defined by the basis $\{v_b\}$ of $V_\mathbb{Q}$. In the case where K has finite characteristic p, this amounts to "reducing mod p" the coefficients of $R_\mathbb{Q}$.

Our notation in the preceding discussion conceals a disadvantage: in general there are many different $\mathbb{Z}$-forms $V_\mathbb{Z}$, $V'_\mathbb{Z}, \ldots$ of a given $\mathbb{Q}\Gamma_\mathbb{Q}$-module $V_\mathbb{Q} \in M_\mathbb{Q}(n,r)$, and the corresponding $K\Gamma_K$-modules $V_K = V_\mathbb{Z} \otimes K$, $V'_K = V'_\mathbb{Z} \otimes K, \ldots$ may be not all isomorphic. However one of the classical results of modular theory, deducible from [5, p. 258, (8)] or [49, p. 44, théorème 2] or [20, p. 162, (2.5a)], says that, for any type of simple $K\Gamma_K$-module $L_\lambda \in M_K(n,r)$, the multiplicity $m_\lambda(V_K)$ of L_λ as a composition factor in V_K depends only on $V_\mathbb{Q}$, i.e. is the same for all $\mathbb{Z}$-forms $V_\mathbb{Z}$ of $V_\mathbb{Q}$.

In the case that $V_{\mathbb{Q}} = V_\mu$ is a simple $\mathbb{Q}\Gamma_{\mathbb{Q}}$-module, this multiplicity is often written $d_{\mu\lambda}$, and referred to as a *decomposition number* for the modular reduction $M_{\mathbb{Q}}(n,r) \to M_K(n,r)$.

2.6 The module $\mathbf{E}^{\otimes \mathbf{r}}$

Fix our infinite field K, and write $\Gamma = \Gamma_K = \mathsf{GL}_n(K)$. Let

$$E = E_K = K \cdot e_1 \oplus \cdots \oplus K \cdot e_n$$

be an n-dimensional K-space with a basis $\{ e_\nu : \nu \in \underline{n} \}$ on which Γ acts "naturally":

$$ge_\nu = \sum_{\mu \in \underline{n}} g_{\mu\nu} e_\mu = \sum_{\mu \in \underline{n}} c_{\mu\nu}(g) e_\mu \,, \ \text{all } g \in \Gamma,\ \nu \in \underline{n}.$$

Since the corresponding invariant matrix is $C = (c_{\mu\nu})$, we see that the $K\Gamma$-module E is an object of $A_K(n,1)$.

Now let $r \geq 1$, then Γ acts on the r-fold tensor power $E^{\otimes r} = E \otimes \cdots \otimes E$ in the usual way ($\otimes$ here means $\otimes_K$). The space $E^{\otimes r}$ has K-basis

$$\{ e_i = e_{i_1} \otimes \cdots \otimes e_{i_r} \ : \ i \in I(n,r) \},$$

and relative to this the action of Γ is given by

$$ge_j = ge_{j_1} \otimes \cdots \otimes ge_{j_r} = \sum_{i \in I(n,r)} g_{i_1 j_1} \cdots g_{i_r j_r} e_i = \sum_{i \in I(n,r)} c_{i,j}(g) e_i \,, \quad \text{all } g \in \Gamma,\ j \in I(n,r).$$

The corresponding invariant matrix is $(c_{i,j}) = C \times \cdots \times C$, and this shows that $E^{\otimes r} \in M_K(n,r)$. According to what was said in 2.4, $E^{\otimes r}$ can be regarded as an $S_K(n,r)$-module, by the rule

(2.6a) $$\xi e_j = \sum_{i \in I(n,r)} \xi(c_{i,j}) e_i, \text{ all } \xi \in S_K(n,r),\ j \in I(n,r).$$

In a very famous paper [48] which appeared in 1927, Schur rederived all the results of his 1901 dissertation [47] by an analysis of this module $E^{\otimes r}$. Although his method gives a complete answer only when $\mathsf{char}\, K = 0$, it is still valuable for fields of finite characteristic. We make the symmetric group $G(r)$, and hence also its group algebra $KG(r)$, act on the right of $E^{\otimes r}$ by

(2.6b) $$e_i \pi = e_{i\pi}, \text{ all } i \in I(n,r),\ \pi \in G(r).$$

It is clear that this action commutes with that of $K\Gamma$, or (what is the same) with that of $S_K(n,r)$; we can verify from (2.6a) that $(\xi x)\pi = \xi(x\pi)$ for all $\xi \in S_K(n,r)$, $x \in E^{\otimes r}$ and $\pi \in G(r)$. We have however a stronger statement.

(2.6c) Theorem (Schur). *Let* $\psi : S_K(n,r) \to \mathsf{End}_K(E^{\otimes r})$ *be the representation afforded by the* $S_K(n,r)$*-module* $E^{\otimes r}$*. Then*

(i) $\mathsf{Im}\,\psi = \mathsf{End}_{KG(r)}(E^{\otimes r})$, *and*

(ii) $\mathsf{Ker}\,\psi = 0$.

Hence $S_K(n,r) \cong \mathsf{End}_{KG(r)}(E^{\otimes r})$.

Proof. Each element $\theta \in \mathsf{End}_K(E^{\otimes r})$ has matrix, say $(T_{i,j})$, relative to the basis $\{e_i\}$ of $E^{\otimes r}$. Here i, j run independently over the set $I = I(n,r)$, of course, and the $T_{i,j} \in K$. From (2.6b) follows at once that θ lies in $\mathsf{End}_{KG(r)}(E^{\otimes r})$ if and only if

(2.6d) $\quad T_{i\pi,j\pi} = T_{i,j}$, for all $i,j \in I$ and all $\pi \in G(r)$.

Consequently $\mathsf{End}_{KG(r)}(E^{\otimes r})$ has a K-basis in one-to-one correspondence with the set Ω of all $G(r)$-orbits on $I \times I$, namely if ω is such an orbit, define the corresponding basis element θ_ω to be that $\theta \in \mathsf{End}_K(E^{\otimes r})$ whose matrix $(T_{i,j})$ has $T_{i,j} = 1$ or 0 according as $(i,j) \in \omega$ or not. Now it follows very readily from (2.6a) that, for any $(p,q) \in I \times I$, the basis element $\xi_{p,q}$ of $S_K(n,r)$ is represented on $E^{\otimes r}$ by $\psi(\xi_{p,q}) = \theta_\omega$, where ω is the $G(r)$-orbit containing (p,q). Therefore ψ induces an isomorphism $S_K(n,r) \to \mathsf{End}_{KG(r)}(E^{\otimes r})$, and this proves the theorem.

Remark. The proof of (2.6c) shows that $S_K(n,r)$ has a faithful matrix representation by the algebra of all $n^r \times n^r$ matrices $(T_{i,j})$ which satisfy condition (2.6d). The basis element $\xi_{p,q}$ is represented by the matrix having $T_{i,j} = 1$ or 0 according as $(i,j) \sim (p,q)$ or not. The idempotents $\xi_{i,i}$ are represented by diagonal matrices, and the "orthogonal" decomposition (2.3d) is easy to deduce from this.

(2.6e) Corollary (Schur [47, 48]). *If* $\mathsf{char}\,K = 0$*, or if* $\mathsf{char}\,K = p > r$*, then* $S_K(n,r)$ *is semisimple. Hence every* $V \in M_K(n,r)$ *is completely reducible.*

Proof. Under the given conditions on $\mathsf{char}\,K$, the group algebra $KG(r)$ is semisimple (since $\mathsf{char}\,K$ does not divide $|G(r)| = r!$). Therefore every $KG(r)$-module, and in particular $E^{\otimes r}$, is completely reducible. But the endomorphism algebra of a completely reducible module is semisimple, so by (2.6c), $S_K(n,r)$ is semisimple. The equivalence of categories $M_K(n,r)$ and $\mathsf{mod}\,S_K(n,r)$ now completes the proof of (2.6e).

The family of modules $(E_K^{\otimes r})$, with r fixed but with K varying, is clearly "defined over $\mathbb{Z}$" in the sense of the following definition (which is simply a version of the definition of GL_n-module, GL_n being regarded as affine group scheme over $\mathbb{Z}$. See [49, p. 46]).

Definition. Suppose that for each infinite field K we have a $K\Gamma_K$-module $V_K \in M_K(n,r)$. We say that the family $\{V_K\}$ is *defined over* $\mathbb{Z}$ if there is a $\mathbb{Z}$-form $V_{\mathbb{Z}}$ of $V_{\mathbb{Q}}$, and for each K an isomorphism $\delta_K : V_{\mathbb{Z}} \otimes K \cong V_K$ in the category $M_K(n,r)$. More exactly we say $\{V_K\}$ is $\mathbb{Z}$*-defined by* $V_{\mathbb{Z}}$ *and* $\{\delta_K\}$.

Example 1. Take $V_K = E_K^{\otimes r}$. The module $V_{\mathbb{Z}} = \sum_{i \in I(n,r)} \mathbb{Z} \cdot e_i$ is a $\mathbb{Z}$-form of $V_{\mathbb{Q}}$ (we write $e_{\mu,K}$, $e_{i,K} = e_{i_1,K} \otimes \cdots \otimes e_{i_r,K}$ for the basis elements of E_K, $E_K^{\otimes r}$), and for each K the K-map $\delta_K : V_{\mathbb{Z}} \otimes K \to V_K$ taking $e_i \otimes 1_K \mapsto e_{i,K}$, for all $i \in I(n,r)$, is an isomorphism in $M_K(n,r)$. So $\{E_K^{\otimes r}\}$ is defined over $\mathbb{Z}$.

Definition. Suppose $\{V_K\}$, $\{W_K\}$ are both families of modules in $M_K(n,r)$, both defined over $\mathbb{Z}$, by $V_{\mathbb{Z}}$ and $\{\delta_K\}$, $W_{\mathbb{Z}}$ and $\{\eta_K\}$, respectively. Suppose we have for each K a morphism $\theta_K : V_K \to W_K$ in $M_K(n,r)$. We say that the family $\{\theta_K\}$ is *defined over* $\mathbb{Z}$ if $\theta_{\mathbb{Q}}$ maps $V_{\mathbb{Z}}$ into $W_{\mathbb{Z}}$, and for each K the diagram shown commutes.

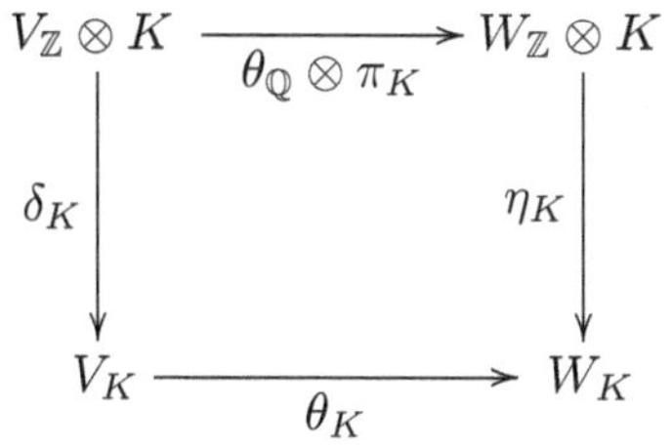

Example 2. Define the rth *symmetric power* $D_{r,K} = D_r(E_K)$ of E_K to be the rth homogeneous subspace of the polynomial ring $K[e_1, \ldots, e_n]$; the elements $e_1 = e_{1,K}, \ldots, e_n = e_{n,K}$ are regarded as commuting indeterminates. There is a surjective K-map $\theta_K : E_K^{\otimes r} \to D_r(E_K)$ taking $e_i = e_{i_1} \otimes \cdots \otimes e_{i_r}$ to the monomial $e_{(i)} = e_{i_1} \cdots e_{i_r}$, for all $i \in I(n,r)$. It is well-known that $D_{r,K}$ has a unique structure as a $K\Gamma$-module, such that θ_K becomes a $K\Gamma$-map; in fact the action on $D_{r,K}$ of a given $g \in \Gamma$, is the restriction to $D_{r,K}$ of the unique K-algebra automorphism of $K[e_1, \ldots, e_n]$ which takes $e_\mu \mapsto g e_\mu$ for all $\mu \in \underline{n}$. We can show that the family $\{D_{r,K}\}$ is defined over $\mathbb{Z}$; the relevant $\mathbb{Z}$-form $D_{r,\mathbb{Z}}$ in $D_{r,\mathbb{Q}}$ is the set of all homogeneous polynomials of degree r in the variables $e_1 = e_{1,\mathbb{Q}}, \ldots, e_n = e_{n,\mathbb{Q}}$, which have coefficients in $\mathbb{Z}$. The isomorphism $\eta_K : D_{r,\mathbb{Z}} \otimes K \to D_{r,K}$ takes $e_{(i),\mathbb{Q}} \otimes 1_K \mapsto e_{(i),K}$ for all $i \in I(n,r)$. It is clear now that the family of morphisms $\{\theta_K\}$ is defined over $\mathbb{Z}$ in the sense of the last definition.

2.7 Contravariant duality

In this section we keep K fixed and write $\Gamma = \Gamma_K$.

The dual space $V^* = \mathsf{Hom}_K(V, K)$ of a $K\Gamma$-module $V \in M_K(n,r)$ can be made into a *right* $K\Gamma$-module in a natural way: if $f \in V^*$, $g \in \Gamma$, we define $fg \in V^*$ by $(fg)(v) = f(gv)$ for all $v \in V$. To make V^* into a *left* $K\Gamma$-module the traditional practice in group theory is to use the map $g \mapsto g^{-1}$ to "reverse" multiplication; one defines a left action (denoted by a dot) of Γ on V^* by $(g \cdot f)(v) = f(g^{-1}v)$. However the $K\Gamma$-module V^* so defined will not, in general, belong to our category $M_K(n,r)$. But if we replace g^{-1} by g^{tr} (transposed matrix) in the above definition, we get a left $K\Gamma$-module structure on V^* which is still in $M_K(n,r)$. We denote by V° the space V^*, equipped with this action

(2.7a) $(g \cdot f)(v) = f(g^{\mathrm{tr}}v)$, all $g \in \Gamma$, $f \in V^*$, $v \in V$.

The module V° is called the "contravariant dual" to V; an analogous dual applies to rational modules over all semisimple algebraic groups Γ, and has been used a great deal in recent years by Wong [56], Verma [53] and Jantzen [29].

It is convenient to express (2.7a) in terms of the action of $S_K(n,r)$. It is easy to see that the K-linear map $J : S_K(n,r) \to S_K(n,r)$, defined by $J(\xi_{i,j}) = \xi_{j,i}$ for all $i,j \in I(n,r)$, is an involutory anti-automorphism of $S_K(n,r)$. In fact one has clearly, for any $\xi \in S_K(n,r)$

(2.7b) $J(\xi)(c_{i,j}) = \xi(c_{j,i})$, all $i,j \in I(n,r)$,

and by taking $\xi = e_g$ we find that $J(e_g) = e_{g^{\mathrm{tr}}}$, all $g \in \Gamma$. So if $V \in M_K(n,r)$ is regarded as $S_K(n,r)$-module the action (2.7a) which defines the contravariant dual V° $(= V^*)$ reads

(2.7c) $(\xi \cdot f)(v) = f(J(\xi)v)$, all $\xi \in S_K(n,r)$, $f \in V^*$, $v \in V$.

It is clear that $V \mapsto V^\circ$ gives an exact contravariant functor on $M_K(n,r)$, and that the usual isomorphism $V \to (V^*)^*$ of K-spaces, gives a natural isomorphism $V \to (V^\circ)^\circ$.

Definition. Let V, W be modules in $M_K(n,r)$. Then a K-bilinear form

$$(\, , \,) : V \times W \to K$$

is called *contravariant* if it has the property.

(2.7d) $(\xi v, w) = (v, J(\xi)w)$, all $\xi \in S_K(n,r)$, $v \in V$, $w \in W$.

The proof of the next proposition is standard.

(2.7e) Proposition. *If $V, W \in M_K(n,r)$ are given, there is a bijective correspondence between contravariant forms $(\, , \,) : V \times W \to K$ and morphisms $\Lambda : V \to W^\circ$ in $M_K(n,r)$, given by*

$$\Lambda(v)(w) = (v,w), \ \textit{all } v \in V, \, w \in W.$$

The form $(\, , \,)$ is non-singular (=non-degenerate) if and only if Λ is an isomorphism.

Example 1. We can use the last proposition to show that the module $E^{\otimes r}$ is self-dual, that is that $E^{\otimes r} \cong (E^{\otimes r})^\circ$. For there is clearly a non-singular bilinear form $\langle\ ,\ \rangle : E^{\otimes r} \times E^{\otimes r} \to K$, defined by $\langle e_i, e_j \rangle = \delta_{ij}$, all $i, j \in I(n,r)$. But a simple calculation, using (2.6a), shows that $\langle\ ,\ \rangle$ satisfies the contravariant condition (2.7d). Call $\langle\ ,\ \rangle$ the *canonical form* on $E^{\otimes r}$.

Example 2. Let $\{V_K\}$ be a family of modules ($V_K \in M_K(n,r)$) which is $\mathbb{Z}$-defined by $V_\mathbb{Z}$ and $\{\delta_K\}$ as in the last section. Let $\{v_{a,\mathbb{Q}}\}$ (a running over some finite index set B) be a basis of $V_\mathbb{Q}$ which $\mathbb{Z}$-generates $V_\mathbb{Z}$. For each K, write $v_{a,K} = \delta_K(v_{a,\mathbb{Q}} \otimes 1_K)$ so that $\{ v_{a,K} : a \in B \}$ is a basis of V_K.

We can now show easily that *the family* $\{V_K^\circ\}$ *is defined over* $\mathbb{Z}$. For each K, write $\{f_{a,K}\}$ for the basis of $V_K^* = V_K^\circ$ dual to $\{v_{a,K}\}$. Then $V_\mathbb{Z}^\circ = \{ f \in V_\mathbb{Q} : f(V_\mathbb{Z}) \subseteq V_\mathbb{Z} \}$ is a $\mathbb{Z}$-form of $V_\mathbb{Q}^\circ$, having $\mathbb{Z}$-basis $\{f_{a,\mathbb{Q}}\}$. It is not hard to see that $\{V_K^\circ\}$ is $\mathbb{Z}$-defined by $V_\mathbb{Z}^\circ$ and the maps $\hat{\delta}_K : V_\mathbb{Z}^\circ \otimes K \to V_K^\circ$ which take $f_{a,\mathbb{Q}} \otimes 1_K \mapsto f_{a,K}$, all $a \in B$.

Example 3. Let $\{V_K\}$ and $\{W_K\}$ be families (V_K and W_K both in $M_K(n,r)$) which are $\mathbb{Z}$-defined by $V_\mathbb{Z}$, $\{\delta_K\}$ and $W_\mathbb{Z}$, $\{\eta_K\}$ respectively. If for every K we have a bilinear form $(\ ,\ )_K : V_K \times W_K \to K$, we say that the family $\{(\ ,\ )_K\}$ is *defined over* $\mathbb{Z}$ if $(\ ,\ )_\mathbb{Q}$ maps $V_\mathbb{Z} \times W_\mathbb{Z}$ to $\mathbb{Z}$, and for each K, and for each $v_\mathbb{Z} \in V_\mathbb{Z}$, $w_\mathbb{Z} \in W_\mathbb{Z}$ there holds

$$\Big(\delta_K(v_\mathbb{Z} \otimes 1_K), \eta_K(w_\mathbb{Z} \otimes 1_K)\Big)_K = (v_\mathbb{Z}, w_\mathbb{Z})_\mathbb{Q} \cdot 1_K .$$

If all the $(\ ,\ )_K$ are contravariant, then we can show that the family of morphisms $\Lambda_K : V_K \to W_K^\circ$ (given, for each K, by Proposition (2.7e)) is defined over $\mathbb{Z}$. For example, the family of canonical forms $\langle\ ,\ \rangle_K$ on $E_K^{\otimes r}$ is defined over $\mathbb{Z}$, and so therefore is the family of isomorphisms $E_K^{\otimes r} \to (E_K^{\otimes r})^\circ$ derived from these forms.

2.8 $A_K(n,r)$ as $K\Gamma$-bimodule

We saw in the introduction (p. 2) that the space K^Γ of all functions $f : \Gamma \to K$ is a $K\Gamma$-bimodule, i.e. it is a left and right $K\Gamma$-module, and these two actions of $K\Gamma$ commute. If we take an element $c \in A_K(n,r)$, then there holds an equation $\Delta(c) = \sum_t c_t \otimes c'_t$, for suitable elements $c_t, c'_t \in A_K(n,r)$ (see 2.3). By formulae (1c), (1d),

(2.8a) $$g \circ c = \sum_t c'_t(g)c_t, \quad c \circ g = \sum_t c_t(g)c'_t,$$

for any $g \in \Gamma$. This shows that $A_K(n,r)$ is closed to both the left and the right $K\Gamma$-actions, hence $A_K(n,r)$ is itself a $K\Gamma$-bimodule. Now extend (2.8a) by linearity, to give the action of an arbitrary element $\kappa \in K\Gamma$:

$$\kappa \circ c = \sum_t c'_t(\kappa) c_t\,, \quad c \circ \kappa = \sum_t c_t(\kappa) c'_t\,.$$

By (2.4b), we find $\kappa \circ c = c \circ \kappa = 0$, for any $\kappa \in Y = \mathsf{Ker}\, e$. Then (2.4c) shows that $A_K(n,r)$, regarded as left $K\Gamma$-module, belongs to $M_K(n,r)$. Similarly $A_K(n,r)$ belongs to an analogously defined category $M'_K(n,r)$ of right $K\Gamma$-modules (see 4.4). Thus $A_K(n,r)$ becomes an $S_K(n,r)$-bimodule, with actions

(2.8b) $\quad \xi \circ c = \sum_t \xi(c'_t) c_t, \quad c \circ \xi = \sum_t \xi(c_t) c'_t,$ for $\xi \in S_K(n,r)$.

Now let us define a bilinear form $(\ ,\ ) : S_K(n,r) \times A_K(n,r) \to K$ by the rule: if $\xi \in S_K(n,r)$, $c \in A_K(n,r)$, then

$$(\xi, c) = J(\xi)(c).$$

This is non-singular, in fact $\{\xi_{i,j}\}$, $\{c_{i,j}\}$ are dual bases with respect to $(\ ,\ )$ (see (2.7b)). The reader may check that for any $\xi, \eta \in S_K(n,r)$, $c \in A_K(n,r)$ there holds

(2.8c) $\quad (\xi\eta, c) = (\eta, J(\xi) \circ c) = (\xi, c \circ J(\eta)).$

In fact, using (2.3a), (2.8b) and the definition of $(\ ,\ )$, we see that all three expressions just given are equal to $\sum_t (\xi, c'_t)\,(\eta, c_t)$. But (2.8c) shows that $(\ ,\ )$ is contravariant, $S_K(n,r)$ and $A_K(n,r)$ being regarded as left $S_K(n,r)$-modules (and even when they are regarded as right modules). By (2.7e) we deduce that $A_K(n,r) \cong (S_K(n,r))^\circ$, an isomorphism of $K\Gamma$-bimodules.

3

Weights and Characters

3.1 Weights

In this chapter we describe the theory of weights and characters of modules in $M_K(n,r)$, in terms of the Schur algebra $S_K(n,r)$. All our results go back to Schur [47], and all have been generalized in classical researches of Weyl and Chevalley. For the generalization to the category of rational representations of reductive group schemes split over a principal ideal domain, see [49, §3].

Let $n, r \geq 1$ be given, and let $\Lambda(n,r)$ denote the set of all $G(r)$-orbits in $I(n,r)$. The elements $\alpha, \beta, \ldots$ of $\Lambda(n,r)$ will be called *weights* (more precisely, they are weights of GL_n, of dimension r). A weight α is specified by the vector $\alpha = (\alpha_1, \ldots, \alpha_n)$ which gives the *content* of any $i = (i_1, \ldots, i_r) \in \alpha$, i.e. for each $\nu \in \underline{n}$, α_ν is the number of $\varrho \in \underline{r}$ such that $i_\varrho = \nu$. These vectors α can also be regarded as unordered partitions of r into n parts (zero parts being allowed).

The symmetric group $W = G(n)$ can be identified with the *Weyl group* of GL_n. It acts on $I(n,r)$ on the left, $wi = (w(i_1), \ldots, w(i_r))$ for any $w \in W$ and $i \in I(n,r)$. This action commutes with the action of $G(r)$ on the right, and therefore W acts on $\Lambda(n,r)$: $w^{-1}\alpha = (\alpha_{w(1)}, \ldots, \alpha_{w(n)})$ for $\alpha \in \Lambda(n,r)$ and $w \in W$. Each W-orbit of $\Lambda(n,r)$ contains exactly one *dominant weight*, i.e. a weight λ such that $\lambda_1 \geq \cdots \geq \lambda_n$. Thus dominant weights correspond to (ordered) partitions of r into not more than n parts. Denote by $\Lambda^+(n,r)$ the set of all dominant weights.

3.2 Weight spaces

Fix an infinite field K. If $i \in I(n,r)$ belongs to the weight $\alpha \in \Lambda(n,r)$ we shall denote the idempotent $\xi_{i,i}$ (see 2.3) by ξ_α. This is reasonable, since $\xi_{i,i} = \xi_{j,j}$ if and only if $i \sim j$. The orthogonal decomposition (2.3d) now reads

$$\varepsilon = \sum_{\alpha \in \Lambda(n,r)} \xi_\alpha \,.$$

If we apply this to any module $V \in M_K(n,r)$ we get

(3.2a) $$V = \bigoplus_{\alpha \in \Lambda(n,r)} \xi_\alpha V,$$

a decomposition of V as a direct sum of subspaces $\xi_\alpha V$.

In fact [47, pp. 6,7], $\xi_\alpha V$ coincides with the *α-weight-space* V^α of V, which is defined as

$$V^\alpha = \{ v \in V \,:\, x(t)v = t_1^{\alpha_1} \cdots t_n^{\alpha_n}\, v, \text{ all } x(t) \in T_n(K) \}.$$

Here $T_n(K)$ is the *diagonal subgroup* (a maximal split torus) of $\Gamma_K = \mathsf{GL}_n(K)$ consisting of all diagonal matrices $x(t) = \mathsf{diag}(t_1, \ldots, t_n)$ with $t_1, \ldots, t_n$ in $K^* = K \backslash \{0\}$. For each $\alpha \in \Lambda = \Lambda(n,r)$, define the multiplicative character $\chi^\alpha : T_n(K) \to K^*$ by $\chi^\alpha(x(t)) = t_1^{\alpha_1} \cdots t_n^{\alpha_n}$.

To show that $\xi_\alpha V = V^\alpha$, first verify the formula

(3.2b) $$e_{x(t)} = \sum_{\alpha \in \Lambda} t_1^{\alpha_1} \cdots t_n^{\alpha_n} \xi_\alpha, \text{ all } x(t) \in T_n(K),$$

by evaluating both sides at each $c_{i,j}$ $(i,j \in I(n,r))$. If $v \in \xi_\alpha V$, we deduce $x(t)v = e_{x(t)}v = t_1^{\alpha_1} \cdots t_n^{\alpha_n}\, v$; hence $\xi_\alpha V \subseteq V^\alpha$. But for distinct elements $\alpha, \beta \in \Lambda(n,r)$ the multiplicative characters χ^α, χ^β are unequal (since K is infinite), and by a familiar argument it follows that the sum of the weight-spaces V^α, $\alpha \in \Lambda(n,r)$, is direct. Comparing this with (3.2a), we see that $\xi_\alpha V = V^\alpha$ for each α:

(3.2c) $$V = \bigoplus_{\alpha \in \Lambda} V^\alpha.$$

Remark. It follows from (3.2b), and the fact that K is infinite, that the image of $K \cdot T_n(K)$ under the map e (see 2.4) is the subalgebra $D_K(n,r)$ of $S_K(n,r)$ which has the ξ_α $(\alpha \in \Lambda(n,r))$ as K-basis. This is a commutative, split, semisimple algebra.

Example. For each r satisfying $0 \le r \le n$, the rth exterior power $V = \Lambda^r E$ $(= \mathsf{Alt}^r(E))$ is a $K\Gamma$-module in $M_K(n,r)$. If $s = \{i_1, \ldots, i_r\}$ is any r-element subset of $\underline{n}$ $(i_1 < i_2 < \cdots < i_r)$ write $e_s = e_{i_1} \wedge \ldots \wedge e_{i_r}$. These $\binom{n}{r}$ elements e_s form a K-basis of V. Moreover if $x(t) \in T_n(K)$ and $\alpha = \alpha(s)$ is the weight containing $(i_1, \ldots, i_r)$, then $x(t)e_s = t_{i_1} \cdots t_{i_r} e_s = t_1^{\alpha_1} \cdots t_n^{\alpha_n} e_s$. Clearly, distinct r-element subsets s, s' of $\underline{n}$ give distinct weights $\alpha(s)$, $\alpha(s')$. So the weight-spaces V^α all have dimension 1 or zero: $V^{\alpha(s)} = K \cdot e_s$ for any r-element subset $s \subseteq \underline{n}$, and $V^\alpha = 0$ for all other $\alpha \in \Lambda(n,r)$.

3.3 Some properties of weight spaces

Let V be a module in $M_K(n,r)$, and α an element of $\Lambda(n,r)$.

(3.3a) Proposition. *Let $w \in W = G(n)$. Then the K-spaces V^α, $V^{w(\alpha)}$ are isomorphic.*

Proof. Let n_w be the element of $\Gamma = \mathsf{GL}_n(K)$ which maps the basis elements $e_1, \ldots, e_n$ of E to $e_{w(1)}, \ldots, e_{w(n)}$ respectively. It is simple to verify

$$n_w^{-1} x(t_1, \ldots, t_n) n_w = x(t_{w(1)}, \ldots, t_{w(n)}),$$

for all $t_1, \ldots, t_n \in K^*$, hence that $v \mapsto n_w v$ gives a K-isomorphism from V^α onto $V^{w(\alpha)}$.

(3.3b) Proposition. *Let*

$$0 \to V_1 \to V \to V_2 \to 0$$

be an exact sequence in $M_K(n,r)$. Then the naturally induced sequence of K-spaces

$$0 \to V_1^\alpha \to V^\alpha \to V_2^\alpha \to 0$$

is exact.

Proof. The second sequence is obtained by applying ξ_α to every term of the first. Since ξ_α is idempotent, the result follows.

Now let r, s be any non-negative integers and V, W be $K\Gamma$-modules belonging to $M_K(n,r)$, $M_K(n,s)$ respectively. Clearly $V \otimes W = V \otimes_K W$, regarded as $K\Gamma$-module in the usual way, belongs to $M_K(n, r+s)$. It is elementary to verify the

(3.3c) Proposition. *Let $\gamma \in \Lambda(n, r+s)$. Then*

$$(V \otimes W)^\gamma = \bigoplus_{\alpha,\beta} V^\alpha \otimes W^\beta,$$

where the sum is over all $\alpha \in \Lambda(n,r)$, $\beta \in \Lambda(n,s)$ such that $\alpha + \beta = \gamma$.

Next suppose L is a field containing K. We identify $S_K(n,r)$ with a subset of $S_L(n,r)$ by identifying $\xi^K_{i,j}$ with $\xi^L_{i,j}$, for all $i,j \in I(n,r)$. Then $\xi^K_\alpha = \xi^L_\alpha$, for all $\alpha \in \Lambda(n,r)$. So if we make $V_L = V \otimes_K L$ into an $S_L(n,r)$-module by "extension of scalars", and identify V with the subset $V \otimes 1_L$ of V_L, we have the

(3.3d) Proposition. *The weight-space $V_L^\alpha = \xi^L_\alpha V_L$ is the L-span of the weight space $V^\alpha = \xi^K_\alpha V$. In particular, $\mathsf{dim}_K V^\alpha = \mathsf{dim}_L V_L^\alpha$.*

Contravariant duality (see 2.7) behaves well with respect to weight-spaces. If $V, W \in M_K(n,r)$ and $(\ ,\ ) : V \times W \to K$ is a contravariant form, then $(V^\alpha, W^\beta) = 0$ for any distinct weights $\alpha, \beta \in \Lambda(n,r)$ (see [56, p. 42], or [29, p. 6]). For we have $J(\xi_\alpha) = \xi_\alpha$, so by the contravariant property $(\xi_\alpha v, \xi_\beta w) = (v, \xi_\alpha \xi_\beta w) = 0$. It follows that $(\ ,\ )$ is non-singular if and only if the restrictions $(\ ,\ )_\alpha : V^\alpha \times W^\alpha \to K$ are non-singular for all $\alpha \in \Lambda(n,r)$. Taking $W = V^\circ$, there follows the

(3.3e) Proposition. $\mathsf{dim}_K V^\alpha = \mathsf{dim}_K (V^\circ)^\alpha$, *for all* $\alpha \in \Lambda(n,r)$.

Finally let $\{V_K\}$ be a family of modules, $\mathbb{Z}$-defined by $V_\mathbb{Z}$, $\{\delta_K\}$ as in 2.6. Because the idempotents $\xi_\alpha^\mathbb{Q}$ in $S_\mathbb{Q}(n,r)$ actually lie in $S_\mathbb{Z}(n,r)$, we have a direct sum $V_\mathbb{Z} = \bigoplus_\alpha \xi_\alpha^\mathbb{Q} V_\mathbb{Z}$, and $\xi_\alpha^\mathbb{Q} V_\mathbb{Z} = V_\mathbb{Z} \cap V_\mathbb{Q}^\alpha$, for all $\alpha \in \Lambda(n,r)$. These $\xi_\alpha^\mathbb{Q} V_\mathbb{Z}$, being summands of the free $\mathbb{Z}$-module $V_\mathbb{Z}$, are themselves free $\mathbb{Z}$-modules. We have the

(3.3f) Proposition. *For each* K, V_K^α *is the* K*-span of the image under the map* $\delta_K : V_\mathbb{Z} \otimes K \to V_K$ *of* $\xi_\alpha^\mathbb{Q} V_\mathbb{Z} \otimes 1_K$. *Hence* $\mathsf{dim}_K V_K^\alpha$ *equals the rank of the free* $\mathbb{Z}$*-module* $\xi_\alpha^\mathbb{Q} V_\mathbb{Z}$, *and so is independent of* K.

3.4 Characters

Let V be a $K\Gamma_K$-module in $M_K(n,r)$. To each weight $\alpha \in \Lambda(n,r)$ we assign the monomial $X_1^{\alpha_1} \cdots X_n^{\alpha_n}$ of degree r in n indeterminates $X_1, \ldots, X_n$ over the rational field $\mathbb{Q}$. Then the *character* (or *formal character*, cf. [32, p. 274]) of V is defined to be the polynomial

$$\Phi_V(X_1, \ldots, X_n) = \sum_{\alpha \in \Lambda(n,r)} (\mathsf{dim}_K V^\alpha) \cdot X_1^{\alpha_1} \cdots X_n^{\alpha_n} .$$

Thus Φ_V is an element of the polynomial ring $\mathbb{Z}[X_1, \ldots, X_n]$, and is homogeneous of degree r. By (3.3a) it is symmetric, in fact

$$\textbf{(3.4a)} \qquad \Phi_V(X_1, \ldots, X_n) = \sum_\lambda (\mathsf{dim}_K V^\lambda) \cdot \mathsf{m}_\lambda(X_1, \ldots, X_n),$$

the sum being over all dominant weights $\lambda \in \Lambda^+(n,r)$. Here m_λ is the *monomial symmetric function* (see for example [39, p. 11]), i.e. the sum of the distinct monomials obtained from $X_1^{\lambda_1} \cdots X_n^{\lambda_n}$ by permuting $X_1, \ldots, X_n$. For example, let r be in the range $0 \le r \le n$, then the character of the exterior power $V = \Lambda^r E$ (see 3.2) is the rth *elementary symmetric function* $\mathsf{e}_r = \mathsf{m}_{(1,1,\ldots,1,0,\ldots,0)} = X_1 X_2 \cdots X_r + \cdots$.

The propositions in section 3.3 give rise to propositions about characters. Suppose $0 \to V_1 \to V \to V_2 \to 0$ is an exact sequence in $M_K(n,r)$, then $\Phi_V = \Phi_{V_1} + \Phi_{V_2}$, by (3.3b). It follows by induction on the length l of a composition series of V, say

$$V = V_0 \supset V_1 \supset V_2 \supset \cdots \supset V_l = 0,$$

that

$$\textbf{(3.4b)} \qquad \Phi_V = \sum_{\sigma=1}^{l} \Phi_{V_{\sigma-1}/V_\sigma} .$$

From (3.3c) we have

(3.4c) $\Phi_{V\otimes W} = \Phi_V \, \Phi_W$

when $V \in M_K(n,r)$, $W \in M_K(n,s)$; this extends of course to a similar formula for tensor products of any finite number of factors. For example if $\mu = (\mu_1, \ldots, \mu_r)$ is any partition of r (i.e. $\mu_1 \geq \cdots \geq \mu_r \geq 0$ and $\mu_1 + \cdots + \mu_r = r$) then the symmetric function

$$\mathsf{e}_\mu(X_1, \ldots, X_n) = \mathsf{e}_{\mu_1} \cdots \mathsf{e}_{\mu_r}$$

is the character of the module $\Lambda^{\mu_1} E \otimes \cdots \otimes \Lambda^{\mu_r} E$. Since every character Φ_V lies in the ring $\mathsf{Sym}(n,r)$ of all symmetric functions in $\mathbb{Z}[X_1, \ldots, X_n]$, and since by the fundamental theorem on symmetric functions ([39, (2.4), p. 13]) $\mathsf{Sym}(n,r)$ is $\mathbb{Z}$-spanned by the e_μ above, we have the well-known

(3.4d) Theorem. *The additive subgroup of* $\mathbb{Z}[X_1, \ldots, X_n]$ *which is generated by all characters* Φ_V, $V \in M_K(n,r)$, *is* $\mathsf{Sym}(n,r)$. *In particular, this additive group is independent of the field* K.

We must next connect our "formal" character Φ_V with the *natural character* φ_V of $V = (V, \varrho) \in M_K(n,r)$, defined by

$$\varphi_V(g) = \mathsf{Trace}\, \varrho(g), \text{ all } g \in \Gamma = \mathsf{GL}_n(K).$$

The map φ_V is an element of $A_K(n,r)$, in fact φ_V is the trace of the invariant matrix (r_{ab}) afforded by any basis $\{v_a\}$ of V.

(3.4e) Theorem (see [47, p. 17]). *Let* $V \in M_K(n,r)$ *and* $g \in \mathsf{GL}_n(K)$, *then* $\varphi_V(g) = \Phi_V(\zeta_1, \ldots, \zeta_n)$, *where* $\zeta_1, \ldots, \zeta_n$ *are the eigenvalues of* g.

Proof. By (3.3d), the character Φ_V is unchanged if we replace V by the module $V_L = V \otimes_K L \in M_L(n,r)$ obtained by extending K to a larger field L. Since this process replaces φ_V by a function on $\Gamma_L = \mathsf{GL}_n(L)$ which coincides with φ_V on $\Gamma_K = \mathsf{GL}_n(K)$, we may legitimately assume, in proving (3.4e), that K is algebraically closed.

Let C be the $n \times n$ matrix $(c_{\mu\nu})$, and let u be an indeterminate over K. Define elements $f_1, \ldots, f_n \in A_K(n,r)$ by

(1) $\det(uI - C) = u^n - f_1 u^{n-1} + \cdots + (-1)^n f_n$.

It is clear that $f_r(g) = \mathsf{e}_r(\zeta_1, \ldots, \zeta_n)$, for $1 \leq r \leq n$. Now we may write

$$\Phi_V = \sum_\mu b_\mu \mathsf{e}_1^{\mu_1} \cdots \mathsf{e}_r^{\mu_r} \quad (b_\mu \in \mathbb{Z}),$$

the sum being over the partitions μ of r, as above. Define the element ψ of $A_K(n,r)$ by $\psi = \sum_\mu (b_\mu \cdot 1_K) f_1^{\mu_1} \cdots f_r^{\mu_r}$. Then we have, for any $g \in \Gamma_K$

(2) $\Phi_V(\zeta_1, \ldots, \zeta_n) = \psi(g)$.

Now suppose g is diagonalizable, i.e. that there is some $z \in \Gamma_K$ such that $zgz^{-1} = \mathsf{diag}(\zeta_1, \ldots, \zeta_n)$. Relative to a basis of V which is adapted to the decomposition $V = \bigoplus V^\alpha$, $\mathsf{diag}(\zeta_1, \ldots, \zeta_n)$ is represented by a diagonal matrix having $\dim V^\alpha$ diagonal terms $\zeta_1^{\alpha_1} \cdots \zeta_n^{\alpha_n}$, for each $\alpha \in V(n,r)$. Taking traces we have

(3) $\varphi_V(g) = \varphi_V(zgz^{-1}) = \Phi_V(\zeta_1, \ldots, \zeta_n)$.

We have now two polynomials in n^2 variables $c_{\mu\nu}$, namely ψ and φ_V, and by (2) and (3), $\psi(g) = \varphi_V(g)$ for all g in the set D of diagonalizable elements of $\Gamma_K = \mathsf{GL}_n(K)$. Since every matrix whose eigenvalues are distinct belongs to D, we have $\psi(g) = \varphi_V(g)$ for all $g \in \Gamma_K$ satisfying $d(g) \neq 0$, where $d \in A_K(n,r)$ is the discriminant of the polynomial (1). It follows that $\psi = \varphi_V$, and this completes the proof of (3.4e).

Corollary. *Suppose that $\Phi_1, \ldots, \Phi_t$ are the characteristics of a set of mutually non-isomorphic, absolutely irreducible modules $V_1, \ldots, V_t \in M_K(n,r)$. Then $\Phi_1, \ldots, \Phi_t$ are linearly independent elements of* $\mathsf{Sym}(n,r)$.

Proof. A theorem of Frobenius-Schur (see [11, p. 184,(27.13)]) shows that the natural characters $\varphi_1, \ldots, \varphi_t$ of $V_1, \ldots, V_t$ are linearly independent elements of $A_K(n,r)$. (It is here that we need *absolute* irreducibility.) If there is a non-trivial relation $z_1\Phi_1 + \cdots + z_t\Phi_t = 0$ ($z_\tau \in \mathbb{Z}$), we may assume, in case $p = \mathsf{char}\, K$ is finite, that p does not divide all the z_τ. Then by (3.4e) $(z_1 \cdot 1_K)\varphi_1(g) + \cdots + (z_t \cdot 1_K)\varphi_t(g) = 0$ for all $g \in \Gamma_K$. But this contradicts the Frobenius-Schur theorem.

3.5 Irreducible modules in $\mathbf{M_K(n,r)}$

The next theorem, forerunner of far-reaching generalizations by Weyl [54] and Chevalley [49], is due in the case $K = \mathbb{C}$ to Schur [47, p. 37].

The leading term of a polynomial in $\mathbb{Z}[X_1, \ldots, X_n]$ is taken relative to the usual lexicographical (Gaussian) ordering of weights $\alpha \in \Lambda(n,r)$, or of monomials $X_1^{\alpha_1} \cdots X_n^{\alpha_n}$ (see [47, p. 17]).

(3.5a) Theorem. *Let n, r be given integers, $n \geq 1$, $r \geq 0$. Let K be an infinite field. Then*

(i) *For each $\lambda \in \Lambda^+(n,r)$ there exists an absolutely irreducible module $F_{\lambda,K}$ in $M_K(n,r)$ whose character $\Phi_{\lambda,K}$ has leading term $X_1^{\lambda_1} \cdots X_n^{\lambda_n}$.*

(ii) *These $\Phi_{\lambda,K}$ ($\lambda \in \Lambda^+(n,r)$) form a $\mathbb{Z}$-basis of* $\mathsf{Sym}(n,r)$.

(iii) *Every irreducible module $V \in M_K(n,r)$ is isomorphic to $F_{\lambda,K}$ for exactly one $\lambda \in \Lambda^+(n,r)$.*

Proof. (i) (see [47, p. 37]) Let $\mu = (\mu_1, \ldots, \mu_r)$ be the partition conjugate to λ. We saw in 3.4 that the module $V = \Lambda^{\mu_1}E \otimes \cdots \otimes \Lambda^{\mu_r}E$ has character $\Phi_V = \mathsf{e}_1^{\mu_1} \cdots \mathsf{e}_r^{\mu_r}$. The leading term of Φ_V is therefore $X_1^{\lambda_1} \cdots X_n^{\lambda_n}$. By (3.4b) there is some composition factor U of V whose character has leading term $X_1^{\lambda_1} \cdots X_n^{\lambda_n}$. We may take U to be $F_{\lambda,K}$. To prove that U is absolutely irreducible, it is enough to show that every $K\Gamma_K$-endomorphism θ of U is scalar (see [11, p. 202, (29.13)]). Since our assumption on Φ_U shows that $\dim U^\lambda = 1$, and since θ must map U^λ into itself, there is some $a \in K$ such that $\theta_U(u) = a \cdot u$ for $u \in U^\lambda$. But the set $U' = \{ u \in U : \theta(u) = a \cdot u \}$ is a submodule of U, hence $U = U'$ and θ is equal to scalar map $a \cdot 1_U$.

(ii) The monomial symmetric functions m_λ ($\lambda \in \Lambda^+(n,r)$) form a basis of $\mathsf{Sym}(n,r)$. If we express the functions $\Phi_{\lambda,K}$ in terms of these, we get equations of the form $\Phi_{\lambda,K} = \mathsf{m}_\lambda + \sum_{\mu<\lambda} z_{\lambda\mu}\mathsf{m}_\mu$, and it follows that the characters $\Phi_{\lambda,K}$ ($\lambda \in \Lambda^+(n,r)$) also form a basis for $\mathsf{Sym}(n,r)$.

(iii) Suppose L is an algebraically closed field containing K. Denote by $F^L_{\lambda,K}$ the module $F_{\lambda,K} \otimes_K L \in M_L(n,r)$. Since $F_{\lambda,K}$ is absolutely irreducible, so is $F^L_{\lambda,K}$. On the other hand $F^L_{\lambda,K}$ has the same character $\Phi_\lambda = \Phi_{\lambda,K}$ as $F_{\lambda,K}$, by (3.3d). Any irreducible $X \in M_L(n,r)$ must be isomorphic to one of the $F^L_{\lambda,K}$, since otherwise the characters Φ_X, Φ_λ ($\lambda \in \Lambda^+(n,r)$) would be linearly independent by Corollary (3.4e), and that would contradict (ii) above.

Now let V be any irreducible module in $M_K(n,r)$, and let X be a minimal submodule of $V_L = V \otimes_K L$. There is some $\lambda \in \Lambda^+(n,r)$, then, such that $F^L_{\lambda,K} \cong X$; hence the space $\mathsf{Hom}_{S_L(n,r)}(F^L_{\lambda,K}, V_L) \neq 0$, it follows $\mathsf{Hom}_{S_K(n,r)}(F_{\lambda,K}, V) \neq 0$. Therefore $V \cong F_{\lambda,K}$, and the proof of Theorem (3.5a) is complete.

Remarks.

(i) The proof above shows that $F_{\lambda,K}$ is defined, uniquely up to isomorphism in $M_K(n,r)$, by the properties

(a) $F_{\lambda,K} \in M_K(n,r)$ is irreducible, and

(b) $F_{\lambda,K}$ has character $\Phi_{\lambda,K}$ whose leading term is $X_1^{\lambda_1} \cdots X_n^{\lambda_n}$.

Moreover if L is any field containing K, then $F_{\lambda,K} \otimes_K L \in M_L(n,r)$ has the corresponding properties (a), (b), since in fact its character is the same as that of $F_{\lambda,K}$. In other words, $\Phi_{\lambda,K} = \Phi_{\lambda,L}$ if K, L are infinite fields with $K \subseteq L$. It follows that $\Phi_{\lambda,K}$ is the same, for all infinite fields K of given characteristic. Write $\Phi_{\lambda,p}$ for $\Phi_{\lambda,K}$, if $p = \mathsf{char}\, K$. For each p, the set $\{ \Phi_{\lambda,p} : \lambda \in \Lambda^+(n,r) \}$ is a $\mathbb{Z}$-basis of $\mathsf{Sym}(n,r)$. If $p > 0$, the coefficients $d_{\lambda\mu}$ which appear in the equations

$$\Phi_{\lambda,0} = \sum_{\mu \in \Lambda^+(n,r)} d_{\lambda\mu}\Phi_{\mu,p}\,, \quad \lambda \in \Lambda^+(n,r)$$

are by definition the *decomposition numbers* relative to the modular reduction from $M_{\mathbb{Q}}(n,r)$ to $M_K(n,r)$, where K is any field of characteristic p (see 2.5).

(ii) Suppose that K is fixed. Let $R(M_K(n))$ be the Grothendieck ring for the category $M_K(n)$, and let $[V]$ be the element of $R(M_K(n))$ corresponding to a module $V \in M_K(n)$. Then it is clear from (3.4a) and what was said in 3.3, that $[V] \mapsto \Phi_V$ defines an isomorphism of rings from $R(M_K(n))$ onto the ring $\mathsf{Sym}(n) = \sum_{r\geq 0} \mathsf{Sym}(n,r)$ of all symmetric functions in $\mathbb{Z}[X_1, \ldots, X_n]$.

(iii) The symmetric functions $\Phi_{\lambda,0}$ were determined by Schur [47, p. 23], and are now known as *Schur functions* or *S-functions* (see [39, p. 24]). For finite characteristic p, the irreducible characters $\Phi_{\lambda,p}$ have not yet been calculated explicitly. We sketch here a proof, rather different from Schur's, of his theorem for characteristic zero. If $\alpha \in \Lambda(n,r)$ we write X^α for $X_1^{\alpha_1} \cdots X_n^{\alpha_n}$; recall that $W = G(n)$ acts on $\Lambda(n,r)$ (see 3.1), and write $\mathsf{s}(w)$ for the "sign" ($\in \{-1, 1\}$) of a permutation $w \in W$. Define $\mathsf{a}_\alpha = \mathsf{a}_\alpha(X_1, \ldots, X_n) = \sum_{w\in W} \mathsf{s}(w) X^{w(\alpha)}$, an alternating function in $X_1, \ldots, X_n$, which in fact is expressible as the $n \times n$ determinant $|X_i^{\alpha_j}|$. Let $\delta = (n-1, n-2, \ldots, 1, 0) \in \Lambda(n, \frac{1}{2}n(n-1))$. Define the S-function $\mathsf{S}_\lambda = \mathsf{S}_\lambda(X_1, \ldots, X_n)$ for any $\lambda \in \Lambda^+(n,r)$ by $\mathsf{S}_\lambda = \mathsf{a}_{\lambda+\delta}/\mathsf{a}_\delta$. Then (see [39, (3.2), (4.3)]) the S_λ form a $\mathbb{Z}$-basis of $\mathsf{Sym}(n,r)$ (in fact S_λ has leading term X^λ), and there holds the formal identity

$$(1)\quad \prod_{\mu,\nu=1}^{n} \frac{1}{1 - X_\mu Y_\nu} = \sum_{\lambda\in\Lambda^+(n)} \mathsf{S}_\lambda(X)\, \mathsf{S}_\mu(Y),$$

where $Y = (Y_1, \ldots, Y_n)$ is a second set of variables, independent of the set $X = (X_1, \ldots, X_n)$, and $\Lambda^+(n) = \bigcup_{r\geq 0} \Lambda^+(n,r)$.

Schur's theorem [47, p. 23] is that $\Phi_{\lambda,0} = \mathsf{S}_\lambda$, for all $\lambda \in \Lambda^+(n,r)$. To prove this, it is enough to show that (1) remains true when S_λ's are replaced by $\Phi_{\lambda,0}$'s. For then, considering only the part of (1) which is of given degree r in (both) X and Y, we shall have

$$(2)\quad \sum_{\lambda\in\Lambda^+(n,r)} \mathsf{S}_\lambda(X)\mathsf{S}_\lambda(Y) = \sum_{\lambda\in\Lambda^+(n,r)} \Phi_{\lambda,0}(X)\Phi_{\lambda,0}(Y).$$

If we write $\mathsf{S}_\lambda = \sum v_{\lambda\mu}\Phi_\mu$ ($\lambda \in \Lambda^+(n,r)$) we get an integral matrix $(v_{\mu\nu})$ which, on account of (2), must be orthogonal. This matrix must therefore be a signed permutation matrix, i.e. the S_λ's coincide with $\Phi_{\lambda,0}$'s up to order and sign (cf. [39, p. 35]). But since both S_λ and $\Phi_{\lambda,0}$ have leading term X^λ, we must have $\mathsf{S}_\lambda = \Phi_{\lambda,0}$ for all $\lambda \in \Lambda^+(n,r)$.

So we must prove (1), with S_λ's replaced by $\Phi_{\lambda,0}$'s. It is clearly sufficient to prove that the sums of terms, of each given degree $r \geq 0$, coincide on the two sides of the proposed equation. So we must prove

$$(3)\quad \sum_{(r)} \Big(\prod_{\mu,\nu=1}^{n} X_\mu^{r_{\mu\nu}} Y_\nu^{r_{\mu\nu}} \Big) = \sum_{\lambda\in\Lambda^+(n)} \Phi_{\lambda,0}(X)\, \Phi_{\lambda,0}(Y),$$

where the sum on the left is over all non-negative integral matrices such that $\sum_{\mu,\nu} r_{\mu\nu} = r$.
Let K be an infinite field of characteristic zero. Because the module $E^{\otimes r}$ is completely reducible (2.6e), and contains each (absolutely) irreducible $F_{\lambda,K}$ ($\lambda \in \Lambda^+(n,r)$) with positive multiplicity ($S_K(n,r)$ acts *faithfully* on $E^{\otimes r}$, see (2.6c)) we have

$$(4)\ A_K(n,r) = \bigoplus_{\lambda \in \Lambda^+(n,r)} \mathsf{cf}(F_{\lambda,K}).$$

Now $A_K(n,r)$ is a $K\Gamma$-bimodule, using right and left translation operators $g \circ c$, $c \circ g$ (see 2.8). Each coefficient space $\mathsf{cf}(F_{\lambda,K})$ is a sub-bimodule of $A_K(n,r)$. Take diagonal matrices $x = \mathsf{diag}(x_1, \ldots, x_n)$ and $y = \mathsf{diag}(y_1, \ldots, y_n)$ ($x_\mu, y_\nu \in K^*$), then calculate the trace $f(x,y)$ of the linear transformation $c \mapsto y \circ c \circ x$ ($c \in A_K(n,r)$).

Using the basis of monomials $c_{i,j}$ for $A_K(n,r)$, we find that $f(x,y)$ is obtained by substituting x for X, y for Y in the left side of (3). But from (4) we get another basis of $A_K(n,r)$, by taking for each $\lambda \in \Lambda^+(n,r)$ the coefficient function r^λ_{ab} appearing in an invariant matrix (r^λ_{ab}) for $F_{\lambda,K}$, relative to some basis of $F_{\lambda,K}$. (The Frobenius-Schur theorem (see 3.4) shows that these functions r^λ_{ab} are linearly independent.) If we choose a basis for $F_{\lambda,K}$ which is adapted to the weight-space decomposition, so that each basis element belongs to some weight-space $F^\alpha_{\lambda,K}$, it is easy to show that the trace of the map $c \mapsto y \circ c \circ x$ ($c \in \mathsf{cf}(F_{\lambda,K})$) is $\Phi_\lambda(X)\Phi_\lambda(Y)$; hence by (4), $f(x,y) = \sum_\lambda \Phi_\lambda(X)\Phi_\lambda(Y)$. But this proves (3), since it holds for arbitrary substitutions $x_1, \ldots, x_n, y_1, \ldots, y_n \in K^*$ of the variables $X_1, \ldots, X_n, Y_1, \ldots, Y_n$.

4

The modules $D_{\lambda,K}$

4.1 Preamble

In this section and the next we shall define, for each $\lambda = (\lambda_1, \ldots, \lambda_n)$ in $\Lambda^+(n,r)$ and for each infinite field K, the modules $D_{\lambda,K}$ and $V_{\lambda,K}$ (see introduction). Both have character $\Phi_{\lambda,0} = \mathsf{S}_\lambda(X_1, \ldots, X_n)$.

In 4.4, we shall define $D_{\lambda,K}$ in terms of certain determinantal functions in $A_K(n,r)$. These modules have been known, in the "classical" case $K = \mathbb{C}$, for a very long time—they were discovered (under the guise of "primary covariants") by J. Deruyts in 1892 [13, p. 72][1]. More recently they have been described, for fields of arbitrary characteristic, by M. Clausen [8, p. 180], and by G. D. James [27, p. 129]. In fact James refers to $D_{\lambda,K}$ as a "Weyl module", but we prefer to reserve this term for the module $V_{\lambda,K}$ which was defined by Carter-Lusztig [6, p. 211], and which, as we show in 5.2, is the contravariant dual of $D_{\lambda,K}$.

However James' construction in [27] gives more information than ours, since it can be used to identify $D_{\lambda,K}$ with the "induced module" of a one-dimensional character on the lower triangular Borel subgroup $B_n^-(K)$ of $\Gamma_K = \mathsf{GL}_n(K)$. We mention this identification in 4.8.

4.2 λ-tableaux

For the rest of §4, $\lambda = (\lambda_1, \ldots, \lambda_n) \in \Lambda^+(n,r)$ is fixed. The *diagram* (or "shape") of λ is defined to be the subset

$$[\lambda] = \{ (s,t) \, : \, 1 \leq s, \, 1 \leq t \leq \lambda_s \}$$

of $\mathbb{Z} \times \mathbb{Z}$ (cf. [15, p. 66]). A λ-*tableau* is a map, not necessarily bijective, of $[\lambda]$ into a set. Since $[\lambda]$ has r elements, there exists at least one bijection $T : [\lambda] \to \underline{r}$. We shall arbitrarily choose one such bijection, and call it

[1] I am indebted to J. Towber for this reference. In his article [52], Towber gives an account of the history of $D_{\lambda,K}$: see particularly [52, p. 448].

the *basic* λ-*tableau* $T = T^\lambda$. If the image under T of (s,t) is $x(s,t)$, we may depict T as

(4.2a)

$$\begin{matrix} x(1,1) & x(1,2) & \cdots & & \cdots & x(1,\lambda_1) \\ x(2,1) & x(2,2) & \cdots & x(2,\lambda_2) & & \\ x(3,1) & x(3,2) & \cdots & & & \\ \cdots & \cdots & & & & \end{matrix}$$

Thus every element $\varrho \in \underline{r}$ appears exactly once in (4.2a). If $\varrho = x(s,t)$, we say that ϱ is in *row* s and *column* t of T. The *row stabilizer* $R(T)$ of T is the subgroup of $G = G(r)$ consisting of all $\pi \in G$ which preserve the rows of (4.2a), and the *column stabilizer* $C(T)$ is defined similarly.

If $i : \underline{r} \to \underline{n}$ is an element of $I(n,r)$, we denote the λ-tableau $i \circ T : [\lambda] \to \underline{n}$ by T_i. In general, T_i is not bijective. If $\varrho = x(s,t)$, we often refer to i_ϱ as the *entry* in place ϱ, or in the (s,t) place, of T_i.

Example. Suppose $r = 5$, $n = 3$ and $\lambda = (3,2,0)$. We might take for our basic λ-tableau $T = \begin{pmatrix} 1 & 3 & 5 \\ 2 & 4 & \end{pmatrix}$. Then, for any $i \in I(n,r)$, $T_i = \begin{pmatrix} i_1 & i_3 & i_5 \\ i_2 & i_4 & \end{pmatrix}$. The entries of T_i belong to the set $\underline{3} = \{1,2,3\}$.

4.3 Bideterminants

Let K be an infinite field, and let i, j be elements of $I(n,r)$. We define an element $(T_i : T_j) = (T_i : T_j)_K$ of $A_K(n,r)$ by the formula

$$\textbf{(4.3a)} \qquad (T_i : T_j) = \sum_{\sigma \in C(T)} \mathsf{s}(\sigma) c_{i,j\sigma} = \sum_{\sigma \in C(T)} \mathsf{s}(\sigma) c_{i\sigma,j}.$$

Here $\mathsf{s}(\sigma)$ denotes the sign of σ. The second equality comes from the fact that $c_{i,j\sigma} = c_{i\sigma^{-1},j}$, for any $\sigma \in G(r)$.

Apart from the interchange of rows and columns, the element $(T_i : T_j)$ is a *bideterminant* in the sense of Désarménien, Kung and Rota [15, p. 67]. Let $\mu = (\mu_1, \ldots, \mu_r)$ be the partition of r conjugate to λ (notice that μ might not be an element of $\Lambda^+(n,r)$, since it might have more than n parts). Then it is easy to see that $(T_i : T_j)$ is the product, over all columns t of $[\lambda]$, of the $\mu_t \times \mu_t$ determinants

$$\mathsf{det}\left(c_{i_{x(s,t)}, j_{x(s',t)}}\right)_{s,s'=1,\ldots,\mu_t}.$$

(If $\mu_t = 0$, we take this determinant to be 1.) From this, or directly from (4.3a), we see that $(T_i : T_j)$ is zero if there are equal entries at any two places in a column of T_i, or of T_j. Also $(T_i : T_{j\sigma}) = (T_{i\sigma} : T_j) = \mathsf{s}(\sigma)(T_i : T_j)$, for all $\sigma \in C(T)$.

Example 1. Let $\lambda = (3,2,0)$, T be as in the example in 4.2. Consider $T_l = \begin{pmatrix} 1 & 1 & 1 \\ 2 & 2 & \end{pmatrix}$, $T_i = \begin{pmatrix} a & d & f \\ b & e & \end{pmatrix}$. Then

$$(T_l : T_i) = \begin{vmatrix} c_{1a} & c_{1b} \\ c_{2a} & c_{2b} \end{vmatrix} \begin{vmatrix} c_{1d} & c_{1e} \\ c_{2d} & c_{2e} \end{vmatrix} c_{1f} \in A_K(3,5).$$

Example 2. Let l be the element of $I(n,r)$ whose λ-tableau is

$$\textbf{(4.3b)} \qquad T_l = \begin{pmatrix} 1 & 1 \cdots & \cdots 1 \\ 2 & 2 \cdots & 2 \\ 3 & 3 \cdots & \\ \cdots & & \end{pmatrix}.$$

In other words, $l_{x(s,t)} = s$, for all $(s,t) \in [\lambda]$. Then

$$(T_l : T_l) = c(\mu_1)c(\mu_2)\cdots,$$

where, for any integer m $(0 \le m \le n)$, $c(m)$ denotes the mth leading minor of the $n \times n$ matrix $C = (c_{\mu\nu})$.

4.4 Definition of $\mathbf{D_{\lambda,K}}$

The space $A_K(n,r)$ is a bimodule for $K\Gamma$, with $\Gamma = \Gamma_K$ acting by right and left translations. Equivalently, it is a bimodule for $S_K(n,r)$. If $h, j \in I(n,r) = I$, we have (see 2.8)

$$\textbf{(4.4a)} \qquad \xi \circ c_{h,j} = \sum_{i \in I} \xi(c_{i,j}) c_{h,i}$$

and

$$\textbf{(4.4a}'\textbf{)} \qquad c_{h,j} \circ \xi = \sum_{i \in I} \xi(c_{h,i}) c_{i,j}\,,$$

for all $\xi \in S_K(n,r)$. As left module $A_K(n,r)$ belongs to the category $M_K(n,r)$, and as right module it belongs to the analogously defined category $M'_K(n,r)$ of all right, finite-dimensional $K\Gamma$- (or $S_K(n,r)$-) modules whose coefficient space lies in $A_K(n,r)$. In fact the coefficient space of $A_K(n,r)$, whether regarded as left or right $K\Gamma$-module, is precisely $A_K(n,r)$.

Now let l be the element of $I(n,r)$ defined in (4.3b). It is clear that l belongs to the weight $\lambda = (\lambda_1, \ldots, \lambda_n)$.

Definition. $D_{\lambda,K}$ is the K-span of all bideterminants $(T_l : T_i)$, $i \in I(n,r)$.

Hence $D_{\lambda,K}$ is a subspace of $A_K(n,r)$. Replace h by $l\sigma$ in (4.4a), multiply by $\mathsf{s}(\sigma)$, and sum over all σ in $C(T)$. We get

(4.4b) $$\xi \circ (T_l : T_j) = \sum_{i \in I} \xi(c_{i,j})(T_l : T_i), \text{ all } \xi \in S_K(n,r).$$

It follows that $D_{\lambda,K}$ is a left $S_K(n,r)$-submodule of $A_K(n,r)$. If we compare (4.4b) with (2.6a), we see also that the surjective linear map

$$\varphi = \varphi_K : E_K^{\otimes r} \to D_{\lambda,K},$$

which takes $e_j \mapsto (T_l : T_j)$ for all $j \in I(n,r)$, is an $S_K(n,r)$-module homomorphism.

Remark. Our definition of l and $(T_l : T_i)$ depends on our choice of basic λ tableau T (see 4.2). However any other bijective λ-tableau T' can be written $T' = \pi T$ for some $\pi \in G(r)$. Then $T'_i = T_{i\pi}$ for any $i \in I(n,r)$, and if $l' = l\pi^{-1}$ we have $(T'_{l'} : T'_i) = (T_l : T_i)$. Therefore $D_{\lambda,K}$ is in fact independent of the choice of T.

Example 1. In case $\lambda = (r, 0, \ldots, 0)$ we shall write $D_{r,K}$ for $D_{\lambda,K}$. The λ diagram is a single row of length r, and $T_l = (1\ 1\ \ldots\ 1)$. For any $i \in I(n,r)$, we have $T_i = (i_1\ i_2\ \ldots\ i_r)$ and $(T_l : T_i) = c_{1,i_1} \cdots c_{1,i_r}$. If we map this to the monomial $e_{i_1} \cdots e_{i_r}$ of 2.6, example 2, we see that $D_{r,K}$ is isomorphic to the rth symmetric power of E.

Example 2. Assume $r \leq n$. In case $\lambda = (1, \ldots, 1, 0, \ldots, 0)$ with r 1's, we shall write $D_{(1^r),K}$ for $D_{\lambda,K}$. The λ diagram is a single column, and T_l has entries $1, 2, \ldots, r$. For any $i \in I(n,r)$, $(T_l : T_i)$ is the r-rowed determinant $\det(c_{l,1})$. Mapping this onto $e_{i_1} \wedge \ldots \wedge e_{i_r}$ (see 3.2), we see that $D_{(1)^r,K}$ is isomorphic to the rth exterior power of $\Lambda^r E_K$.

4.5 The basis theorem for $\mathbf{D_{\lambda,K}}$

As usual, K is an infinite field. Our aim is to prove the

(4.5a) Basis theorem. *$D_{\lambda,K}$ has K-basis consisting of all $(T_l : T_i)$ such that T_i is "standard", i.e. the entries in each row of T_i are weakly increasing ($\leq$) from left to right, and the entries in each column are strictly increasing ($<$) from top to bottom.*

This theorem generalizes the theorem of Specht-Garnir for Specht modules (see [45] and [27, §8, 13]). It may be deduced from a more general basis theorem of Désarménien, Kung and Rota [15, p. 78]. For completeness, and also to prepare for the transition to the module $V_{\lambda,K}$ in §5, we give here a proof of (4.5a) based on a combinatorial lemma of Carter-Lusztig (see 4.6).

We begin by showing that

(4.5b) *The set $\{ (T_l : T_i) : T_i \text{ standard}\}$ is linearly independent.*

Proof. All the $(T_l : T_i)$ $(i \in I(n,r))$ lie in the "right" λ-weight-space

$$^{\lambda}A_K(n,r) = A_K(n,r) \circ \xi_\lambda.$$

For it is clear from (2.3c) that $^{\lambda}A_K(n,r)$ is spanned by the elements $c_{l,i}$, where $i \in I(n,r)$ (remember $\xi_\lambda = \xi_{l,l}$). Now $c_{l,i} = c_{l,j}$ if and only if there is some $\pi \in G(r)$ such that $l\pi = l$ and $i\pi = j$. The condition $l\pi = l$ is equivalent to $\pi \in R(T)$, the row stabilizer of the basic λ-tableau ((4.2a)). So $c_{l,i} = c_{l,j}$ if and only if T_i, T_j can be obtained from one another by a row permutation (thus the $c_{l,i}$ correspond to James' "λ-tabloids" [27, pp. 10, 127]).

For each $i \in I(n,r)$, define $\beta(i) = (\beta_1(i), \ldots, \beta_n(i))$, where $\beta_s(i)$ is the sum of the entries in row s of T_i. Clearly $\beta(i) = \beta(j)$ if $c_{l,i} = c_{l,j}$, by what we have just said; hence

(4.5c) *If* $i, j \in I(n,r)$ *and* $\beta(i) \neq \beta(j)$, *then* $c_{l,i} \neq c_{l,j}$.

We order these vectors $\beta(i)$ (they are in fact elements of $\Lambda(n,r)$) in the usual lexicographic way. The reader may verify

(4.5d) *If* T_i *is standard and if* $1 \neq \pi \in C(T)$, *then* $\beta(i\pi) > \beta(i)$.

Now suppose we have a non-trivial linear relation

$$\sum_{i \in H} f_i\,(T_l : T_i) = 0, \quad f_i \in K,\ f_i \neq 0,$$

where the sum is over a non-empty subset H of those $i \in I(n,r)$ such that T_i is standard. By (4.3a), this gives

$$\textbf{(4.5e)} \qquad \sum_{i \in H} \sum_{\pi \in C(T)} f_i\,\mathsf{s}(\pi)\,c_{l,i\pi} = 0.$$

By (4.5c) we can equate to zero the partial sum of all terms (i, π) in (4.5e) such that $\beta(i\pi)$ is equal to any given $\beta \in \Lambda(n,r)$. Take for β, the least of the $\beta(i\pi)$ $(i \in H,\ \pi \in C(T))$. By (4.5d), we can have $\beta = \beta(i\pi)$, for $i \in H$, $\pi \in C(T)$, only if $\pi = 1$. So the partial sum has the form

$$\textbf{(4.5f)} \qquad \sum_{i \in H'} f_i\,c_{l,i} = 0,$$

for some non-empty subset H' of H. But since T_i is standard for all $i \in H'$, the $c_{l,i}$ appearing in (4.5f) are linearly independent, and this gives a contradiction. This proves (4.5b).

4.6 The Carter-Lusztig lemma

To complete the proof of the basis theorem (4.5a) we must show that every bideterminant $(T_l : T_i)$ $(i \in I(n,r))$ is expressible as a linear combination of $(T_l : T_i)$ for which T_i is standard. This is the harder part of the proof of (4.5a), and we deduce it from a combinatorial lemma (4.6a) of Carter-Lusztig. The proof of (4.6a) is given in [6, p. 214, 215], and does not depend essentially on other results in [6]. The reader may note some variations of expression between Carter-Lusztig's paper and this one: their "semi-standard" is our "standard"; their T corresponds to our T_i; their σT is our $T_{i\sigma}$; the "type" λ' of T $(= T_i)$ corresponds to the "weight" of i. Finally, in our (4.6a) we have *not* restricted f to tableaux of a given type λ'.

(4.6a) Carter-Lusztig lemma. *Let $f : I(n,r) \to F$ be any map with values in an abelian group F, satisfying the following three conditions:*

(i) *$f(i) = 0$, if T_i has equal entries at two distinct places in the same column.*

(ii) *$f(i\sigma) = \mathsf{s}(\sigma)\, f(i)$, for any $i \in I(n,r)$ and $\sigma \in C(T)$.*

(iii) *(Garnir relations) $\sum_{\nu \in G(J)} \mathsf{s}(\nu)\, f(i\nu) = 0$, for any $i \in I(n,r)$ and any non-empty subset J of the $(h+1)^{th}$ column C_{h+1} of the basic λ-tableau T. Here h is any element of $\{1, 2 \ldots, r-1\}$, and $G(J)$ is a transversal of the set of cosets $\{\, \nu X : \nu \in Y \,\}$, where Y is the subgroup of $G(r)$ consisting of all $\pi \in G(r)$ which fix every element outside $\mathsf{C}_h \cup J$, and $X = C(T) \cap Y$.*

Then $\mathsf{Im}\, f$ lies in the subgroup of F generated by the set $\{\, f(i) : T_i$ standard$\}$.

Remarks. Condition (ii) shows that the sum in (iii) is independent of the choice of transversal $G(J)$. Condition (iii) is equivalent, by an argument given in [6, p. 212], to condition (37) of [6, p. 214]. A slightly weaker form of (4.6a), which uses a larger set of "Garnir relations" (but is still adequate for our purposes) can be proved by an adaptation of the proof of Garnir [19] for Specht modules; see [45, pp. 93, 94] or [27, pp. 29, 30].

Lemma (4.6a) completes the proof of the basis theorem (4.5a), as soon as we observe

(4.6b) *The function $f(i) = (T_l : T_i)$ satisfies conditions (i), (ii) and (iii) of (4.6a).*

Proof. That $f(i)$ satisfies (i) and (ii) has already been said in 4.3. The proof of (iii) is like that for Specht polynomials, and goes as follows. Every element of the set $B = Y \cdot C(T)$ has unique expression $\pi = \nu\sigma$, with $\nu \in G(J)$ and $\sigma \in C(T)$. So the left side of the Garnir relation (iii) (with $f(i) = (T_l : T_i)$) becomes

$$\sum_{\nu \in G(J)} \sum_{\sigma \in C(T)} \mathsf{s}(\nu)\, \mathsf{s}(\sigma)\, c_{l,i\nu\sigma} = \sum_{\pi \in B} \mathsf{s}(\pi)\, c_{l,i\pi}\,.$$

An argument given in [45, pp. 92, 93] shows that B can be written as disjoint union of subsets $\{\pi, \pi\kappa\}$, in which κ (which depends on π) is some transposition in $R(T)$. For such a pair

$$\mathsf{s}(\pi)\, c_{l,i\pi} + \mathsf{s}(\pi\kappa)\, c_{l,i\pi\kappa} = 0,$$

because $c_{l,i\pi\kappa} = c_{l\kappa,i\pi} = c_{l,i\pi}$ (notice $l\kappa = l$ for all $\kappa \in R(T)$). Hence the sum above is zero, as required.

4.7 Some consequences of the basis theorem

Let $\alpha \in \Lambda(n,r)$ be a given weight, and suppose $a \in I(n,r)$ is an element of α. Putting ξ_α $(= \xi_{a,a})$ for ξ in formula (4.4b) we see that $\xi_\alpha \circ (T_l : T_j) = (T_l : T_j)$ or zero, according as $j \in \alpha$ or not. Consequently each bideterminant $(T_l : T_j)$ (where $j \in I(n,r)$) is a "weight element" of $D_{\lambda,K}$, i.e. it belongs to the weight-space $D^\alpha_{\lambda,K}$, where α is the weight containing j. Then (4.5a) gives the first statement below:

(4.7a) *For each $\alpha \in \Lambda(n,r)$, $D^\alpha_{\lambda,K}$ has K-basis consisting of all $(T_l : T_i)$ with $i \in \alpha$ and T_i standard. Hence the character of $D_{\lambda,K}$ is equal to the Schur function $\mathsf{S}_\lambda(X_1,\ldots,X_n)$.*

The second statement in (4.7a) follows from the first, and from the fact that the coefficient of X^α in $\mathsf{S}_\lambda = \mathsf{S}_\lambda(X_1,\ldots,X_n)$ is precisely the number of λ-tableaux T_i which have "content" (i.e. weight) α—this can be proved by a direct combinatorial argument from the definition of S_λ (see [39, p. 42]).

Since S_λ is the character of an irreducible module in $M_K(n,r)$ when $\mathsf{char}\,K$ is zero, we deduce

(4.7b) *If $\mathsf{char}\,K = 0$, then $D_{\lambda,K}$ is irreducible.*

We show next that the family $D_{\lambda,K}$ is defined over $\mathbb{Z}$. For each infinite field K, we may write $(T_l : T_i)_K$ to denote the element $(T_l : T_i)$ defined in 4.3.

(4.7c) *The $\mathbb{Z}$-span $D_{\lambda,\mathbb{Z}}$ of the elements $(T_l : T_i)_\mathbb{Q}$ $(i \in I(n,r))$ is a $\mathbb{Z}$-form of $D_{\lambda,\mathbb{Q}}$. The family $\{D_{\lambda,K}\}$ is $\mathbb{Z}$-defined by $D_{\lambda,\mathbb{Z}}$ and the maps*

$$\delta_K : D_{\lambda,\mathbb{Z}} \otimes K \to D_{\lambda,K},\ (T_l : T_i)_\mathbb{Q} \otimes 1_K \mapsto (T_l : T_i)_K$$

Moreover the family of inclusions $D_{\lambda,K} \hookrightarrow {}^\lambda A_K(n,r)$ is also defined over $\mathbb{Z}$.

Proof. When applied to the function $f(i) = (T_l : T_i)_\mathbb{Q}$, the Carter-Lusztig lemma (4.6a) shows that $(T_l : T_i)_\mathbb{Q}$ is in the $\mathbb{Z}$-span of

$$\{\,(T_l : T_i)_\mathbb{Q}\ :\ T_i \text{ standard}\,\}.$$

But by (4.5a), this set is a basis of $D_{\lambda,\mathbb{Q}}$. By (4.4b), $D_{\lambda,\mathbb{Z}}$ is invariant to the left action of $S_\mathbb{Z}(n,r)$. It follows that $D_{\lambda,\mathbb{Z}}$ is a $\mathbb{Z}$-form of $D_{\lambda,\mathbb{Q}}$. The maps δ_K described above are clearly $S_K(n,r)$-morphisms, and are isomorphisms by (4.5a) applied to $D_{\lambda,K}$. The last statement of (4.7c) is immediate from the definitions.

Remark. $D_{\lambda,\mathbb{Z}}$ is a direct summand, as $\mathbb{Z}$-module, of $A_\mathbb{Z}(n,r)$; equivalently, the exact sequence $0 \to D_{\lambda,\mathbb{Z}} \overset{\text{inc}}{\to} A_\mathbb{Z}(n,r)$ is $\mathbb{Z}$-split. This follows from (4.7c) and the next lemma.

(4.7d) Lemma. *Suppose $\{V_K\}$, $\{W_K\}$ are families of modules in $M_K(n,r)$, and are $\mathbb{Z}$-defined by $V_\mathbb{Z}$ and $\{\delta_K\}$, $W_\mathbb{Z}$ and $\{\eta_K\}$. Suppose $\theta_K : V_K \to W_K$ is a family of morphisms in $M_K(n,r)$, also defined over $\mathbb{Z}$ (see 2.6). Then*

(i) *If all the θ_K are injective, then the exact sequence $0 \to V_\mathbb{Z} \stackrel{\theta_\mathbb{Q}}{\to} W_\mathbb{Z}$ is $\mathbb{Z}$-split.*

(ii) *If all the θ_K are surjective, then the exact sequence $V_\mathbb{Z} \stackrel{\theta_\mathbb{Q}}{\to} W_\mathbb{Z} \to 0$ is $\mathbb{Z}$-split.*

Proof. Let $M = (m_{\beta\alpha})$ be the matrix of $\theta_\mathbb{Q}$, relative to bases $\{v_\alpha\}$, $\{w_\beta\}$ of $V_\mathbb{Q}$, $W_\mathbb{Q}$ which $\mathbb{Z}$-generate $V_\mathbb{Z}$, $W_\mathbb{Z}$ respectively. The assumption that the family $\{\theta_K\}$ is defined over $\mathbb{Z}$ implies M is an integral matrix, and that for each K, $M_K = (m_{\beta\alpha} \cdot 1_K)$ is the matrix of θ_K relative to the appropriate bases of V_K, W_K. If all the θ_K are injective, this means that M has rank $d = \dim_\mathbb{Q} V_\mathbb{Q}$, and that it still has rank d, even after reduction modulo any rational prime p. Hence all elementary divisors of M are equal to 1, and therefore there exists an integral matrix $M' = (m'_{\alpha\beta})$ such that $MM' =$ identity. This proves (i). The proof of (ii) is similar.

4.8 James's construction of $\mathbf{D_{\lambda,K}}$

G. D. James has given [27, p. 129] a construction of a $K\Gamma_K$-module W^λ, which he calls a "Weyl module", but which is in fact isomorphic to $D_{\lambda,K}$. If $\alpha, \beta \in \Lambda(n,r)$, then the elements $c_{i,j}$ ($i \in \alpha$, $j \in \beta$) of $A_K(n,r)$ may be identified with James's "α-tabloids of type β" [27, p. 127]. (It is not necessary that either $\alpha = (\alpha_1, \dots, \alpha_n)$ or $\beta = (\beta_1, \dots, \beta_n)$ be dominant, i.e. be proper partitions of r.) James's space $S^{\circ,\lambda}$ becomes ${}^\lambda A_K(n,r)$ in our language (see 4.5). He defines, for all pairs of integers $s \in \{1, \dots, n\}$ and $v \in \{0, \dots, \lambda_{s+1}\}$ a linear map

$$\psi_{s,v} : {}^\lambda A_K(n,r) \to A_K(n,r)$$

by the rule:

(4.8a) $$\psi_{s,v}(c_{l,i}) = \sum_h c_{h,i},$$

summed over all $h \in I(n,r)$ obtained by replacing, in l (or in T_l) $\lambda_{s+1} - v$ of the entries $s+1$ by s.

Definition (see [27, p. 129]). W^λ is the set of all elements $c \in {}^\lambda A_K(n,r)$ such that $\psi_{s,v}(c) = 0$ for all $s \in \{1, \dots, n\}$ and $v \in \{0, \dots, \lambda_{s+1} - 1\}$.

By a rather delicate combinatorial argument, James proves a theorem (see [27, 26.3, p. 128]) from which it follows readily that W^λ has as basis the set $\{(T_l : T_i) : T_i \text{ standard}\}$; hence $W^\lambda = D_{\lambda,K}$. However James's definition has a very important group-theoretical interpretation, which we shall now give.

Let $U^- = U_n^-(K)$ be the subgroup of $\Gamma = \mathsf{GL}_n(K)$ consisting of all lower triangular unipotent matrices in Γ. It is well-known that U^- is generated by the elements $u_s(t)$ ($s \in \{1, \ldots, n-1\}$, $t \in K$) where

$$u_s(t) = \begin{pmatrix} 1 & & & & \\ & \ddots & & & \\ & & 1 & & \\ & & t & 1 & \\ & & & & \ddots & \\ & & & & & 1 \end{pmatrix} \begin{matrix} \\ \\ (\text{row } s) \\ (\text{row } s+1) \\ \\ \\ \end{matrix}$$

with t in position $(s+1, s)$. Now write $g = u_s(t)$ and $i \in I(n,r)$. By definition of the action of Γ on $A_K(n)$ (see introduction)

(4.8b) $$c_{l,i} \circ g = \sum_{h \in I(n,r)} c_{l,h}(g)\, c_{h,i}.$$

It is clear that $c_{l,h}(g) = g_{l_1 h_1} \cdots g_{l_r h_r}$ is zero unless

(4.8c) *For all* $\varrho \in \underline{r}$, $(l_\varrho, h_\varrho) \in \{(1,1), \ldots, (n,n), (s+1, s)\}$,

and if (4.8c) is satisfied, then $c_{l,h}(g) = t^w$, where w is the number of ϱ such that $(l_\varrho, h_\varrho) = (s+1, s)$. In other words, $c_{l,h}(g) = t^w$, for any $h \in I(n,r)$ which can be obtained by replacing, in l (or T_l), w of the entries $s+1$ by s. Thus (4.8b) gives a formula

(4.8d) $$c \circ u_s(t) = \sum_{v=0}^{\lambda_{s+1}} t^{\lambda_{s+1} - v}\, \psi_{s,v}(c),$$

for all $c \in {}^\lambda A_K(n,r)$, $s \in \{1, \ldots, n-1\}$ and $t \in K$. Naturally we prove (4.8d) by taking first the case $c = c_{l,i}$ ($i \in I(n,r)$).

If $v = \lambda_{s+1}$, it is clear that $\psi_{s,v}(c) = c$, for all $c \in {}^\lambda A_K(n,r)$. So by (4.8d), and using the fact that K is infinite, we see that for any $c \in {}^\lambda A_K(n,r)$, the conditions

$$\psi_{s,v}(c) = 0, \text{ all } s \in \{1, \ldots, n-1\},\ v \in \{0, \ldots, \lambda_{s+1} - 1\}$$

are equivalent to the conditions

$$c \circ u_s(t) = c, \text{ all } s \in \{1, \ldots, n-1\},\ t \in K,$$

which in turn are equivalent to the conditions

(4.8e) $$c \circ u = c, \text{ all } u \in U_n^-(K).$$

Recall from 4.5 that ${}^{\lambda}A_K(n,r)$ is the right λ-weight-space of $A_K(n,r)$. Extend the character χ^λ of $T_n(K)$ (see 3.2) to the Borel subgroup

$$B_n^-(K) = T_n(K)\,U_n^-(K),$$

by defining $\chi^\lambda(u) = 1$ for all $u \in U_n^-(K)$. This allows us to reformulate James's theorem as follows.

(4.8f) Theorem. *The module* $D_{\lambda,K} = W^\lambda$ *is the set of all* $c \in A_K(n,r)$ *which satisfy the conditions*

$$c \circ b = \chi^\lambda(b)c, \quad \textit{all } b \in B_n^-(K).$$

This shows that $D_{\lambda,K}$ is the *induced module* $\mathsf{Ind}_{B^-}^{\Gamma}(K_\lambda)$ in the category $M_K(n,r)$ (or, in fact, in the category of all rational modules for $\mathsf{GL}_n(K)$), of a $K \cdot B_n^-(K)$-module K_λ which affords the character χ^λ of $B_n^-(K)$. For a discussion of a theorem equivalent to (4.8f), for semisimple algebraic groups, see [30, §1].

5

The Carter-Lusztig modules $\mathbf{V_{\lambda,K}}$

5.1 Definition of $\mathbf{V_{\lambda,K}}$

Let $\lambda \in \Lambda^+(n,r)$ be given, and let K be any infinite field. Denote by N_K the kernel of the $S_K(n,r)$-epimorphism $\varphi_K : E_K^{\otimes r} \to D_{\lambda,K}$ defined in 4.4. We have then an exact sequence in $M_K(n,r)$

(5.1a) $\quad 0 \longrightarrow N_K \longrightarrow E_K^{\otimes r} \xrightarrow{\varphi_K} D_{\lambda,K} \longrightarrow 0.$

Definition. Let $V_{\lambda,K}$ be the orthogonal complement to N_K, relative to the canonical form $\langle\ ,\ \rangle$ on $E_K^{\otimes r}$ (see 2.7, example 1):

(5.1b) $\quad V_{\lambda,K} = \{ x \in E_K^{\otimes r} : \langle x, N_K \rangle = 0 \}.$

Since $\langle\ ,\ \rangle$ is contravariant and N_K is an $S_K(n,r)$-submodule of $E_K^{\otimes r}$, $V_{\lambda,K}$ is also a submodule of $E_K^{\otimes r}$. Since $\langle\ ,\ \rangle$ is non-singular, we may define a non-singular, contravariant form $(\ ,\ ) : V_{\lambda,K} \times D_{\lambda,K} \to K$ by

$$(x, \varphi_K(y)) = \langle x, y \rangle, \text{ all } x \in V_{\lambda,K},\ y \in E_K^{\otimes r}.$$

Hence $V_{\lambda,K} \cong D^\circ_{\lambda,K}$, and because (contravariant) dual modules in $M_K(n,r)$ have the same character (by (3.3e)), $\Phi_{V_{\lambda,K}} = \mathsf{S}_\lambda(X_1, \ldots, X_n)$. In particular, if $\mathsf{char}\, K = 0$, then $V_{\lambda,K}$ is irreducible, and is isomorphic to $D_{\lambda,K}$ (see 4.7). If K has finite characteristic, then in general $V_{\lambda,K}$ and $D_{\lambda,K}$ are not isomorphic.

5.2 $\mathbf{V_{\lambda,K}}$ is Carter-Lusztig's "Weyl module"

We shall identify $V_{\lambda,K}$ with the module $\overline{V}^\lambda$ defined by Carter and Lusztig in [6, pp. 211, 222]. First we must describe $N_K = \mathsf{Ker}\, \varphi_K$ more exactly.

(5.2a) *N_K is the K-span of the subset $\mathcal{R} = \mathcal{R}_1 \cup \mathcal{R}_2 \cup \mathcal{R}_3$ of $E_K^{\otimes r}$, where*

(i) $\mathcal{R}_1$ *consists of all* e_i *such that* $i \in I(n,r)$ *and* T_i *has equal entries in two different places in some column.*
(ii) $\mathcal{R}_2$ *consists of all* $e_i - \mathsf{s}(\sigma)e_{i\sigma}$, *where* $i \in I(n,r)$ *and* $\sigma \in C(T)$.
(iii) $\mathcal{R}_3$ *consists of all elements* $\sum_{\nu \in G(J)} \mathsf{s}(\nu)\, e_{i\nu}$, *where* $i \in I(n,r)$ *and* J *is a non-empty subset of* $\mathsf{C}_{h+1}(T)$ *for some* $h \in \{1, 2 \ldots, r-1\}$.

Proof. All the elements of $\mathcal{R}$ lie in $N = N_K$ by (4.6b), and so N contains the K-span N' of $\mathcal{R}$. There is therefore a well-defined K-map ψ from $F = E_K^{\otimes r}/N'$ onto $D_{\lambda,K}$, given by $\psi(x + N') = \varphi_K(x)$, all $x \in E_K^{\otimes r}$. Now the function $f : I(n,r) \to F$ given by $f(i) = e_i + N'$ clearly satisfies the hypotheses of (4.6a), hence F is K-spanned by the set $\{\, e_i + N' : T_i \text{ standard} \,\}$. But ψ maps this set onto a basis of $D_{\lambda,K}$, by (4.5a). Therefore $\mathsf{Ker}\,\psi = 0$, which implies $N \subseteq N'$; hence $N = N'$, and (5.2a) is proved.

As a corollary, we have

(5.2b) $V_{\lambda,K}$ *is the set of all elements* $x \in E_K^{\otimes r}$ *which satisfy the following conditions:*

(i) $\langle\, x, e_i \,\rangle = 0$ *for all* $i \in I(n,r)$ *such that* T_i *has equal entries in two distinct places in the same column.*
(ii) $x\sigma = \mathsf{s}(\sigma)\, x$ *for all* $\sigma \in C(T)$.
(iii) $\sum_{\nu \in G(J)} \mathsf{s}(\nu)\, x\nu^{-1} = 0$ *for any* $h \in \{1, 2 \ldots, r-1\}$ *and any non-empty subset* J *of* $\mathsf{C}_{h+1}(T)$.

Proof. (5.2a) shows that $V_{\lambda,K}$ consists of all $x \in E_K^{\otimes r}$ such that $\langle\, x, \mathcal{R}_s \,\rangle = 0$ for $s = 1, 2, 3$. (5.2b) is an almost immediate consequence of this, together with the fact that $\langle\ ,\ \rangle$ is non-singular, and satisfies an "invariance" condition

(5.2c) $\quad \langle\, x\pi, y \,\rangle = \langle\, x, y\pi^{-1} \,\rangle$

for all $x, y \in E_K^{\otimes r}$, $\pi \in G(r)$. This last condition is verified trivially from the definition of $\langle\ ,\ \rangle$ (see 2.7, example 1).

It is now easy to see that $V_{\lambda,K}$ coincides with Carter-Lusztig's "Weyl module" $\overline{V}^{\lambda}$; conditions (28), (29) of [6, p. 211] are essentially (i), (ii), (iii) of (5.2b).

Example 1. If $\lambda = (r, 0, \ldots, 0)$ we write $V_{r,K}$ for $V_{\lambda,K}$. Conditions (i) and (ii) of (5.2b) are vacuous. Condition (iii) says $x\nu = x$, for all transpositions $\nu = (h, h+1)$ $(h = 1, \ldots, r-1)$. So $V_{\lambda,K}$ is the space of all *symmetric tensors* x in $E_K^{\otimes r}$. Therefore this module is dual in the contravariant sense to the rth symmetric power $D_{r,K}$ of E_K (see 4.4, example 1). $V_{r,K}$ has basis $\{\, v_\alpha : \alpha \in \Lambda(n,r) \,\}$, where $v_\alpha = \sum e_i$, sum over all $i \in \alpha$. The non-singular contravariant form $(\ ,\ ) : V_{r,K} \times D_{r,K} \to K$ (see 5.1) is given by $(v_\alpha, e^\beta) = \delta_{\alpha,\beta}$ for all $\alpha, \beta \in \Lambda(n,r)$. Here $e^\beta = e_1^{\beta_1} \cdots e_n^{\beta_n}$, so that $\{\, e^\beta : \beta \in \Lambda(n,r) \,\}$ is a basis of $D_{r,K}$.

Example 2. Assume $r \leq n$ and that $\lambda = (1,\ldots,1,0,\ldots,0)$ with r 1's. We write $V_{(1^r),K}$ for $V_{\lambda,K}$. Condition (iii) of (5.2b) is vacuous, and conditions (i), (ii) show that $V_{(1^r),K}$ is the space of all *antisymmetric tensors* in $E_K^{\otimes r}$. $V_{(1^r),K}$ is an irreducible module in $M_K(n,r)$, whatever the characteristic of K, since its character $\mathsf{e}_r(X_1,\ldots,X_n)$ has no non-trivial expression as a sum of symmetric functions in $\mathbb{Z}[X_1,\ldots,X_n]$. Therefore $V_{(1^r),K}$ is isomorphic to the rth exterior power $D_{(1^r),K} = \Lambda^r E_K$ (see 4.4, example 2). There is an isomorphism $\Lambda^r E_K \to V_{(1^r),K}$ which takes (see 3.3)

$$e_{i_1} \wedge \ldots \wedge e_{i_r} \longmapsto (e_{i_1} \otimes \cdots \otimes e_{i_r})\Big(\sum_{\pi \in G} \mathsf{s}(\pi)\,\pi\Big).$$

5.3 The Carter-Lusztig basis for $\mathbf{V_{\lambda,K}}$

Carter-Lusztig have given a basis for $V_{\lambda,K}$, and shown that $V_{\lambda,K}$ is a cyclic module [6, pp. 216–219]. We shall give here a slightly different proof of these results. It is interesting that the Carter-Lusztig basis (see (5.3b)) is *not* the dual of the basis of $D_{\lambda,K}$ given in (4.5a); these bases of $V_{\lambda,K}$ are connected by a certain unimodular matrix Ω (see (5.3d)) which has appeared in work of Désarménien [14, p. 74].

Notation. If X is any subset of the symmetric group $G(r)$, we shall write

$$[X] = \sum_{\pi \in X} \pi, \qquad \{X\} = \sum_{\pi \in X} \mathsf{s}(\pi)\,\pi.$$

These are elements of the group ring $KG(r)$, of course. If $K = \mathbb{Q}$, they even belong to $\mathbb{Z}G(r)$.

(5.3a) *Let l be the element of $I(n,r)$ given in (4.3b). Then $f_l = e_l\{C(T)\}$ lies in $V_{\lambda,K}$.*

Proof. Verify that $\langle f_l, y \rangle = 0$, for all $y \in \mathcal{R}_1 \cup \mathcal{R}_2 \cup \mathcal{R}_3$ (see (5.2a)). For $y \in \mathcal{R}_1 \cup \mathcal{R}_2$ there is no problem. Suppose then that $y = \sum_{\nu \in G(J)} \mathsf{s}(\nu)\, e_{i\nu}$, as in (5.2a)(iii). In the notation just introduced, this reads $y = e_l\{G(J)\}$. Using (5.2c),

$$\langle e_l\{C(T)\}, e_i\{G(J)\} \rangle = \langle e_l, e_i\{B\} \rangle,$$

where $B = G(J)C(T) = Y \cdot C(T)$—see the proof of (4.6b). As in that proof, we break up B as a union of pairs $\{\pi, \pi\kappa\}$, each κ being a transposition in $R(T)$. Using (5.2c) again, and the fact that $e_l\kappa = e_l$, we have for each such pair $\langle e_l, e_i(\mathsf{s}(\pi)\pi + \mathsf{s}(\pi\kappa)\pi\kappa) \rangle = 0$. Therefore $\langle f_l, y \rangle = 0$, and this property completes the proof of (5.3a).

Remark. The element f_l is denoted by Φ^μ in [6, p. 216].

Now let V' be the $S_K(n,r)$-submodule of $V_{\lambda,K}$ which is generated by f_l. As K-space, V' is spanned by the elements $\xi_{i,j}f_l$ $(i,j \in I(n,r))$. However, since $\xi_{i,j}f_l = (\xi_{i,j}e_l)\{C(T)\}$, we see by (2.6a) that $\xi_{i,j}f_l = 0$ unless $j \sim l$. Hence V' is K-spanned by the elements $b_i = \xi_{i,l}f_l$ $(i \in I(n,r))$. In fact we have the following much more precise statement.

(5.3b) Theorem (see [6, Theorem 3.5, p. 218]). *The set*

$$\{\, b_i = \xi_{i,l}f_l \,:\, i \in I(n,r), T_i \text{ standard}\}$$

is a K-basis of $V_{\lambda,K}$. In particular, $V' = V_{\lambda,K}$, i.e. $V_{\lambda,K}$ is generated by f_l as $S_K(n,r)$- (or $K\Gamma_K$-)module.

Proof. Let $i,j \in I(n,r)$. From (2.6a) we have

(5.3c) $$\xi_{i,l}e_l = \sum_h e_h,$$

where the sum is over all $h \in I(n,r)$ such that $(i,l) \sim (h,l)$, i.e. such that there is some $\pi \in G(r)$ with $h = i\pi$, $l = l\pi$. But $l = l\pi$ if and only if $\pi \in R(T)$ (see (4.3b)). So the sum in (5.3c) is over the $R(T)$-orbit $iR(T)$ of i.

Now we bring in the form $(\ ,\ ) : V_{\lambda,K} \times D_{\lambda,K} \to K$ introduced in 5.1, and calculate

$$\begin{aligned}
(b_i, (T_l : T_j)) &= \langle\, b_i, e_j \,\rangle \\
&= \langle\, \xi_{i,l}e_l\{C(T)\}, e_j \,\rangle \\
&= \langle\, \xi_{i,l}e_l, e_j\{C(T)\} \,\rangle, \quad \text{using (5.2c)} \\
&= \Big\langle \sum_{h \in iR(T)} e_h, \sum_{\sigma \in C(T)} \mathsf{s}(\sigma)e_{j\sigma} \Big\rangle, \quad \text{using (5.3c).}
\end{aligned}$$

This last expression, which we denote $\Omega(i,j)$, is equal to

(5.3d) $$\Omega(i,j) = \sum_\sigma \mathsf{s}(\sigma), \text{ sum over all } \sigma \in C(T) \text{ such that } j\sigma \text{ and } i \text{ belong to the same } R(T)\text{-orbit.}$$

Now suppose T_i, T_j are both standard, and that $\Omega(i,j) \neq 0$. There must exist $\sigma \in C(T)$ such that $j\sigma$ and i are in the same $R(T)$-orbit. If $\sigma = 1$, this implies $i = j$. In any case we have $\beta(i) = \beta(j\sigma)$ (see 4.5), and if $\sigma \neq 1$ then (4.5d) implies $\beta(j\sigma) > \beta(j)$. Let Ω denote the matrix $(\Omega(i,j))$, with i and j running over the set $I^* = \{\, k \in I(n,r) : T_k \text{ standard}\,\}$. If we give I^* any total order $>$ such that $\beta(i) > \beta(j)$ implies $i > j$ for all $i,j \in I^*$, then Ω is a unimodular triangular matrix. For by what we have shown above, $\Omega(i,j) \neq 0$ implies either $i > j$ or $i = j$, and clearly $\Omega(i,i) = 1$. Hence Ω is non-singular. But since $\Omega(i,j) = (b_i, (T_l : T_j))$, and since $(\ ,\ )$ is non-singular and $\{\, (T_l : T_j) : T_j \text{ standard}\,\}$ is a basis of $D_{\lambda,K}$, it follows that $\{\, b_i : T_i \text{ standard}\,\}$ is a basis of $V_{\lambda,K}$. This proves (5.3b).

Note. A proof that Ω is unimodular is given by Désarménien in [14, p. 74].

5.4 Some consequences of the basis theorem

For each $i \in I(n,r)$, it is clear by (2.3c) that the element $b_i = \xi_{i,l} f_l$ satisfies $\xi_{i,i} b_i = b_i$; thus $b_i \in V^{\alpha}_{\lambda,K}$, where α is the weight of i. From (5.3b) we deduce

(5.4a) *Let $\alpha \in \Lambda(n,r)$. Then $V^{\alpha}_{\lambda,K}$ has K-basis $\{ b_i : i \in \alpha,\ T_i \text{ standard} \}$.*

In particular, $V^{\lambda}_{\lambda,K} = K \cdot f_l$ (since $b_l = \xi_{l,l} f_l = f_l$). A well-known argument (see e.g. [31, p. 2]) shows

(5.4b) *$V_{\lambda,K}$ has a unique maximal submodule $V^{max}_{\lambda,K}$. The element f_l does not lie in $V^{max}_{\lambda,K}$. The irreducible module $F_{\lambda,K} = V_{\lambda,K}/V^{max}_{\lambda,K}$ has character $\Phi_{\lambda,p}$, where $p = \mathsf{char}\, K$.*

Proof. By (5.3b) $V_{\lambda,K}$ is generated by f_l. Any proper submodule M of $V_{\lambda,K}$, since it does not contain f_l, has $M^{\lambda} = M \cap V^{\lambda}_{\lambda,K} = 0$, and so M lies in

$$V' = \sum_{\alpha \neq \lambda} V^{\alpha}_{\lambda,K},$$

a proper K-subspace of $V_{\lambda,K}$. Therefore the sum of *all* proper submodules M of $V_{\lambda,K}$ lies in V', hence is proper, and is the unique maximal submodule $V^{\max}_{\lambda,K}$, and does not contain f_l. The third statement in (5.4b) now follows from the definition of $\Phi_{\lambda,p}$ (see 3.5, Remark (i)) and the fact that $V_{\lambda,K}$ has character $\mathsf{S}_{\lambda} = X_1^{\lambda_1} \cdots X_n^{\lambda_n} + \cdots$.

Since $D_{\lambda,K}$ is dual to $V_{\lambda,K}$, we have the following corollary to (5.4b).

(5.4c) *$D_{\lambda,K}$ has a unique minimal submodule $D^{min}_{\lambda,K}$, and $D^{min}_{\lambda,K} \cong (F_{\lambda,K})^{\circ}$. Hence $D^{min}_{\lambda,K}$, $F_{\lambda,K}$ are isomorphic modules.*

Proof. The second statement follows from the first, and the fact that any module V in $M_K(n,r)$ has the same character as its (contravariant) dual V° (see (3.3e)). If V is irreducible this implies $V \cong V^{\circ}$.

Remark. Since $F_{\lambda,K}$ has λ-weight-space of dimension 1 (by (5.4b)) the same is true of $D^{\min}_{\lambda,K}$. Therefore $D^{\min}_{\lambda,K}$ contains, hence is generated as $S_K(n,r)$-module by, the element $(T_l : T_l)$. For (4.7a) shows that the λ-weight-space of $D_{\lambda,K}$ is $K \cdot (T_l : T_l)$. This proves (5.4d) below. A quite different proof comes from a standard argument for semisimple algebraic groups, using the fact (cf. 4.8) that $(T_l : T_l)$ is stable under the action of upper and lower unipotent triangular subgroups of $\Gamma = \mathsf{GL}_n(K)$. See [50, p. 214].

(5.4d) *The element $(T_l : T_l)$ of $D_{\lambda,K}$ generates the irreducible module $D^{min}_{\lambda,K}$.*

We show next that the family $V_{\lambda,K}$ is defined over $\mathbb{Z}$. Recall from 5.1 that $V_{\lambda,\mathbb{Q}} = \{ x \in E_{\mathbb{Q}}^{\otimes r} : \langle x, N_{\mathbb{Q}} \rangle = 0 \}$, where $N_{\mathbb{Q}} = \mathsf{Ker}\, \varphi_{\mathbb{Q}}$.

(5.4e) Lemma. *$V_{\lambda,\mathbb{Z}} = E_{\mathbb{Z}}^{\otimes r} \cap V_{\lambda,\mathbb{Q}}$ is a $\mathbb{Z}$-form of $V_{\lambda,\mathbb{Q}}$. It has $\mathbb{Z}$-basis*

$$B = \{ b_{i,\mathbb{Q}} : i \in I(n,r),\ T_i \text{ standard} \}.$$

Proof. The sets $E_{\mathbb{Z}}^{\otimes r}$, $V_{\lambda,\mathbb{Q}}$ are both closed to the action of $S_{\mathbb{Z}}(n,r)$, hence so is $V_{\lambda,\mathbb{Z}}$. It is clear that, for each $i \in I(n,r)$, $b_{i,\mathbb{Q}} = \xi_{i,l} e_{l,\mathbb{Q}}\{C(T)\}$ lies in $E_{\mathbb{Z}}^{\otimes r}$, hence that B is a subset of $V_{\lambda,\mathbb{Z}}$. Since B is a $\mathbb{Q}$-basis of $V_{\lambda,\mathbb{Q}}$ (by (5.3b)), the proof of (5.4e) will be achieved if we show that every element $x \in V_{\lambda,\mathbb{Z}}$ is in the $\mathbb{Z}$-span of B. We certainly have $x = \sum k_i b_i$ (sum is over T_i standard) for some $k_i \in \mathbb{Q}$. Since $x \in E_{\mathbb{Z}}^{\otimes r}$, we have $\langle x, e_j \rangle \in \mathbb{Z}$ for all $j \in I(n,r)$. Take any j such that T_j is standard. The calculation in the proof of (5.3b) gives $\langle x, e_j \rangle = \sum k_i \Omega(i,j)$, the sum being over the standard T_i. But the "Désarménien matrix" $\Omega = (\Omega(i,j))$, in case $K = \mathbb{Q}$, is integral and unimodular. So from $\sum k_i \Omega(i,j) \in \mathbb{Z}$ (all standard T_j) follows $k_i \in \mathbb{Z}$ (all standard T_i). Hence (5.4e) is proved.

Remark. (5.4e) shows that $V_{\lambda,\mathbb{Z}} = S_{\mathbb{Z}}(n,r) f_l$.

It is clear that $V_{\lambda,\mathbb{Z}} = E_{\mathbb{Z}}^{\otimes r} \cap V_{\lambda,\mathbb{Q}}$ is a pure $\mathbb{Z}$-submodule of $E_{\mathbb{Z}}^{\otimes r}$; hence that the inclusion $0 \to V_{\lambda,\mathbb{Z}} \to E_{\mathbb{Z}}^{\otimes r}$ is $\mathbb{Z}$-split. If we tensor with any infinite field K we get an exact sequence $0 \to V_{\lambda,\mathbb{Z}} \otimes K \to E_{\mathbb{Z}}^{\otimes r} \otimes K$ in the category $M_K(n,r)$. We shall regard $V_{\lambda,\mathbb{Z}} \otimes K$ as submodule of $E_{\mathbb{Z}}^{\otimes r} \otimes K$. In 2.6, example 1 we showed that the family $\{E_K^{\otimes r}\}$ is $\mathbb{Z}$-defined by $E_{\mathbb{Z}}^{\otimes r}$ and maps $\delta_K : e_{i,\mathbb{Q}} \otimes 1_K \mapsto e_{i,K}$. From (5.4e) and (5.3b) it is immediate that δ_K induces an isomorphism

$$\delta'_K : V_{\lambda,\mathbb{Z}} \otimes K \to V_{\lambda,K},$$

which maps $b_{i,\mathbb{Q}} \otimes 1_K \mapsto b_{i,K}$ for all $i \in I(n,r)$. From all this we deduce:

(5.4f) *The family $\{V_{\lambda,K}\}$ is $\mathbb{Z}$-defined by $V_{\lambda,\mathbb{Z}}$ and the maps δ'_K just described. The family of inclusions $V_{\lambda,K} \to E_K^{\otimes r}$ is also defined over $\mathbb{Z}$. So is the family of contravariant forms $(\ ,\ )_K : V_{\lambda,K} \times D_{\lambda,K} \to K$ defined in 5.1.*

5.5 Contravariant forms on $V_{\lambda,K}$

J.C. Jantzen (see [29], [32], ...) has studied contravariant forms on the Weyl modules for a simply-connected, semisimple algebraic group; in particular, his results apply to the group $\mathsf{SL}_n(K)$, and extend with little alteration to our case $\Gamma = \mathsf{GL}_n(K)$. In this section we shall give an independent description of the contravariant forms on the modules $V_{\lambda,K}$.

We saw that $V_{\lambda,K}$ is generated as S $(= S_K(n,r))$-module by the element $f_l = e_l\{C(T)\}$ of $E^{\otimes r}$, and it follows that $V_{\lambda,K}$ is contained in the submodule $E^{\otimes r}\{C(T)\}$ of $E^{\otimes r}$. We have the canonical contravariant form $\langle\ ,\ \rangle$ on $E^{\otimes r}$, and from (5.2c) we deduce that

(5.5a) $\langle x\{C(T)\}, y \rangle = \langle x, y\{C(T)\} \rangle$ *for all* $x, y \in E^{\otimes r}$.

This allows us to define a "contracted" version of $\langle\ ,\ \rangle$ on $E^{\otimes r}\{C(T)\}$ by the rule

(5.5b) *If* $x, y \in E^{\otimes r}$, *define* $\langle\langle x\{C(T)\}, y\{C(T)\} \rangle\rangle = \langle x, y\{C(T)\} \rangle$.

Any ambiguity arising from the fact that an element of $E^{\otimes r}\{C(T)\}$ may be expressed as $x\{C(T)\} = x'\{C(T)\}$ for distinct elements x, x' of $E^{\otimes r}$, is eliminated by (5.5a). It is clear that $\langle\langle\ ,\ \rangle\rangle$ is a symmetric, contravariant form on $E^{\otimes r}\{C(T)\}$. If we restrict it to $V_{\lambda,K}$, we get a symmetric, contravariant form on $V_{\lambda,K}$, which is moreover non-zero, since rule (5.5b) gives

$$\langle\langle f_l, f_l \rangle\rangle = \langle e_l, e_l\{C(T)\} \rangle = \sum_{\sigma \in C(T)} \mathsf{s}(\sigma) \langle e_l, e_{l\sigma} \rangle = 1.$$

We might also mention that the family of forms $\langle\langle\ ,\ \rangle\rangle_K$, constructed in this way for all infinite fields K, is defined over $\mathbb{Z}$ in the sense of 2.7, example 3.

Any contravariant form $(\ ,\ )$ on $V_{\lambda,K}$ coincides, up to a scalar factor, with $\langle\langle\ ,\ \rangle\rangle$. For the contravariant property (2.7d), together with the fact that $V_{\lambda,K} = Sf_l$, shows that $(\ ,\ )$ is determined by the values (f_l, v), $v \in V_{\lambda,K}$. If $v \in V_{\lambda,K}$ is decomposed as a sum $v = \sum v_\alpha$, with each v_α belonging to the weight-space $V^\alpha_{\lambda,K}$ ($\alpha \in \Lambda(n,r)$), then since weight-spaces for distinct weights are orthogonal with respect to $(\ ,\ )$ (see 3.3, p. 25), we have $(f_l, v) = (f_l, v_\lambda)$, i.e. $(\ ,\ )$ is determined by the values (f_l, v) for all v in the λ-weight-space $V^\lambda_{\lambda,K}$. But this weight-space is $K \cdot f_l$ (see (5.4a)). So $(\ ,\ )$ is completely determined by (f_l, f_l). Therefore (since $\langle\langle f_l, f_l \rangle\rangle = 1$) if $(f_l, f_l) = k$, then $(v, w) = k\langle\langle v, w \rangle\rangle$, for all $v, w \in V_{\lambda,K}$.

In the work of Jantzen which we have mentioned, and also in the earlier work of W.J. Wong [56, 57], the importance of the contravariant form on a Weyl module V is that it provides a method of calculating the maximal submodule $V^{\max}$ of V. In our case the result reads as follows.

(5.5c) [57, Theorem 3B, p. 362]. *The radical of* $\langle\langle\ ,\ \rangle\rangle$, *that is, the space* $M = \{ v \in V_{\lambda,K} : \langle\langle v, V_{\lambda,K} \rangle\rangle = 0 \}$, *coincides with the unique maximal submodule* $V^{\max}_{\lambda,K}$ *of* $V_{\lambda,K}$.

Proof. Notice that M is a submodule of $V_{\lambda,K}$, by the contravariant property. Also $M \neq V_{\lambda,K}$, since $f_l \notin M$. Therefore $M \subseteq V^{\max}_{\lambda,K}$. But we saw in the proof of (5.4b) that $V^{\max}_{\lambda,K}$ lies in the sum V' of all weight-spaces $V^\alpha_{\lambda,K}$ ($\alpha \neq \lambda$). Since V' is orthogonal to $V^\lambda_{\lambda,K} = K \cdot f_l$, we have

$$(V^{\max}_{\lambda,K}, V_{\lambda,K}) = \langle\langle V^{\max}_{\lambda,K}, Sf_l \rangle\rangle \subseteq \langle\langle SV^{\max}_{\lambda,K}, f_l \rangle\rangle \subseteq \langle\langle V', f_l \rangle\rangle = 0.$$

This shows that $V^{\max}_{\lambda,K}$ lies in M, and the proof of (5.5c) is complete.

Example. We shall calculate the form $\langle\langle\ ,\ \rangle\rangle$ on $V_{r,K}$ (notation of 5.2, example 1). Since $\lambda = (r, 0, \ldots, 0)$, the diagram of λ has only one row, and so $C(T) = \{1\}$. Therefore, $\langle\langle\ ,\ \rangle\rangle$ is just the restriction to $V_{r,K}$ of the canonical form $\langle\ ,\ \rangle$ on $E^{\otimes r}$. Relative to the basis $\{ v_\alpha : \alpha \in \Lambda(n,r) \}$ given in 5.2, example 1, the form is given by $\langle v_\alpha, v_\beta \rangle = 0$ $(\alpha \neq \beta)$ and $\langle v_\alpha, v_\alpha \rangle = (r,\alpha)\cdot 1_K$, where

$$(r,\alpha) = \frac{r!}{\alpha_1! \cdots \alpha_n!}\,.$$

So the radical M of this form is spanned by those v_α for which $p = \mathsf{char}\, K$ divides the integer (r,α).

Since $M = V_{r,K}^{\max}$, the irreducible module $F_{r,K} = V_{r,K}/V_{r,K}^{\max}$ (see (5.4b)) has basis

$$\{ v_\alpha + M : \alpha \in \Lambda(n,r),\ (r,\alpha) \not\equiv 0 \text{ modulo } p \}.$$

The α-weight-space of $F_{r,K}$ is $K\cdot(v_\alpha + M)$, for all $\alpha \in \Lambda(n,r)$. Hence the character of $F_{r,K}$ is $\Phi_{r,p} = \sum X_1^{\alpha_1} \cdots X_n^{\alpha_n}$, sum over all $\alpha \in \Lambda(n,r)$ with $(r,\alpha) \not\equiv 0$ modulo p. Since the integers (r,α) are the coefficients in the multinomial expansion

$$\textbf{(5.5d)} \qquad (X_1 + \cdots + X_n)^r = \sum_{\alpha \in \Lambda(n,r)} (r,\alpha)\, X_1^{\alpha_1} \cdots X_n^{\alpha_n},$$

we have the result: $\Phi_{r,p}$ (which is a polynomial over $\mathbb{Z}$, by definition) is the sum of those monomials $X_1^{\alpha_1} \cdots X_n^{\alpha_n}$ which have non-zero coefficients when (5.5d) is reduced modulo p. The reader may deduce from this a special case of Steinberg's "Tensor Product Theorem" [50, p. 218]:

If $0 \le r_0, r_1, \ldots \le p-1$ *such that* $r = r_0 + r_1 p + \cdots$, *then*

$$\Phi_{r,p}(X_1, \ldots, X_n) = \prod_{i \ge 0} \Phi_{r_i,p}(X_1^{p^i}, \ldots, X_n^{p^i}).$$

Of course in case $p = 0$, $M = 0$ and we can take $F_{r,0} = V_{r,K}$. The character is the "complete symmetric function" $\mathsf{h}_r = \sum_{\alpha \in \Lambda(n,r)} X^\alpha$ (see [39, p. 14]).

5.6 $\mathbb{Z}$-forms of $\mathbf{V}_{\lambda,\mathbf{K}}$

In this section we work over the rational field $\mathbb{Q}$. We have seen in 5.1 that the modules $V_{\lambda,\mathbb{Q}}$ and $D_{\lambda,\mathbb{Q}}$ are irreducible, and isomorphic to each other. In fact the map $\varphi_\mathbb{Q} : E_\mathbb{Q}^{\otimes r} \to D_{\lambda,\mathbb{Q}}$ induces a map $\varphi : V_{\lambda,\mathbb{Q}} \to D_{\lambda,\mathbb{Q}}$ which is an isomorphism. For φ is certainly a homomorphism, and it is non-zero because

$$\textbf{(5.6a)} \qquad \varphi(f_l) = \sum_{\sigma \in C(T)} \mathsf{s}(\sigma)\, \varphi(e_{l\sigma}) = \sum_{\sigma \in C(T)} \mathsf{s}(\sigma)\, (T_l : T_{l\sigma}) = |C(T)|\, (T_l : T_l).$$

Therefore by Schur's lemma φ is an isomorphism.

We would like to describe all the $\mathbb{Z}$-forms lying in $V_{\lambda,\mathbb{Q}}$. If L is any such $\mathbb{Z}$-form, then the argument at the end of 3.3 shows that $L^\lambda = L \cap V^\lambda_{\lambda,\mathbb{Q}}$ is a free $\mathbb{Z}$-submodule of rank 1 (since $V^\lambda_{\lambda,\mathbb{Q}} = \mathbb{Q} \cdot f_l$ has dimension 1), so that $L^\lambda = \mathbb{Z} \cdot y f_l$, for some $0 \neq y \in \mathbb{Q}$. It is clear that $y^{-1}L$ is also a $\mathbb{Z}$-form of $V_{\lambda,\mathbb{Q}}$, and $(y^{-1}L)^\lambda = \mathbb{Z} \cdot f_l$. Therefore we shall lose nothing essential if we confine our attention to $\mathbb{Z}$-forms L of $V_{\lambda,\mathbb{Q}}$ which are "normalized" by the condition

(5.6b) $\quad L^\lambda = \mathbb{Z} \cdot f_l.$

We already know two such normalized $\mathbb{Z}$-forms, namely

$$V_{\lambda,\mathbb{Z}} = E^{\otimes r} \cap V_{\lambda,\mathbb{Q}} = S_\mathbb{Z}(n,r) \cdot f_l$$

(see (5.4e)), and

(5.6c) $\quad X_{\lambda,\mathbb{Z}} = \varphi^{-1}(\,|C(T)|\, D_{\lambda,\mathbb{Z}}).$

$X_{\lambda,\mathbb{Z}}$ is a $\mathbb{Z}$-form of $V_{\lambda,\mathbb{Q}}$, because $D_{\lambda,\mathbb{Z}}$—hence also $|C(T)|\, D_{\lambda,\mathbb{Z}}$—is a $\mathbb{Z}$-form of $D_{\lambda,\mathbb{Q}}$. It is normalized, because

$$X^\lambda_{\lambda,\mathbb{Z}} = \varphi^{-1}(\,|C(T)|\, D^\lambda_{\lambda,\mathbb{Z}}) = \varphi^{-1}(\mathbb{Z} \cdot |C(T)| \cdot (T_l : T_l)) = \mathbb{Z} \cdot f_l\,,$$

by (4.7c) and (5.6a). Our aim is the following theorem.

(5.6d) (cf. [53, p. 681]). *Let L be any $\mathbb{Z}$-form of $V_{\lambda,\mathbb{Q}}$ which satisfies (5.6b). Then $V_{\lambda,\mathbb{Z}} \subseteq L \subseteq X_{\lambda,\mathbb{Z}}$.*

Proof. We write $S_\mathbb{Z} = S_\mathbb{Z}(n,r)$. Since L contains f_l, it contains $S_\mathbb{Z} \cdot f_l = V_{\lambda,\mathbb{Z}}$. On the other hand $\langle\langle L, V_{\lambda,\mathbb{Z}} \rangle\rangle = \langle\langle L, S_\mathbb{Z} \cdot f_l \rangle\rangle = \langle\langle S_\mathbb{Z} L, f_l \rangle\rangle \subseteq \langle\langle L, f_l \rangle\rangle$, using the contravariant property of $\langle\langle\ ,\ \rangle\rangle$ and the fact that $S_\mathbb{Z} L \subseteq L$. But we know that f_l is orthogonal to all the weight spaces L^α for $\alpha \neq \lambda$. It follows that $\langle\langle L, f_l \rangle\rangle = \langle\langle L^\lambda, f_l \rangle\rangle = \langle\langle \mathbb{Z} \cdot f_l, f_l \rangle\rangle = \mathbb{Z} \cdot \langle\langle f_l, f_l \rangle\rangle = \mathbb{Z}$. We have hereby proved that L lies in the set

$$Y_{\lambda,\mathbb{Z}} = \{ y \in V_{\lambda,\mathbb{Q}} \,:\, \langle\langle y, V_{\lambda,\mathbb{Z}} \rangle\rangle \subseteq \mathbb{Z} \}.$$

So it will be enough to prove that $Y_{\lambda,\mathbb{Z}} = X_{\lambda,\mathbb{Z}}$. That $X_{\lambda,\mathbb{Z}} \subseteq Y_{\lambda,\mathbb{Z}}$ follows from the argument just given, taking $L = X_{\lambda,\mathbb{Z}}$. Conversely let z be any element of $Y_{\lambda,\mathbb{Z}}$. Using the basis theorem (4.5a) for $D_{\lambda,\mathbb{Q}}$ we may write

(5.6e) $$\varphi(z) = \sum_j k_j\, |C(T)|\, (T_l : T_j),$$

the sum being over $j \in I(n,r)$ such that T_j is standard; the k_j lie in $\mathbb{Q}$. Because $z \in Y_{\lambda,\mathbb{Z}}$ we have

$$\langle\langle b_i, z \rangle\rangle = \langle\langle z, b_i \rangle\rangle \in \mathbb{Z}, \text{ for all } i \in I(n,r).$$

Our definition (5.5b) gives a particularly simple formula for $\langle\langle\ ,\ \rangle\rangle$ in the case $\mathsf{char}\, K = 0$, namely

(5.6f) $$\langle\langle\, u, v\,\rangle\rangle = \frac{1}{|C(T)|}\langle\, u, v\,\rangle, \text{ for all } u, v \in E^{\otimes r}\{C(T)\}.$$

For if $u = x\{C(T)\}$, $v = y\{C(T)\}$ as in (5.5b), then $\langle\, u, v\,\rangle = \langle\, x, y\,\{C(T)\}^2\,\rangle$ by (5.2c), and also $\{C(T)\}^2 = |C(T)|\,\{C(T)\}$. We use (5.6f) to make the following calculation:

(5.6g) $$\langle\langle\, b_i, z\,\rangle\rangle = \sum_j k_j \langle\, b_i, \varphi^{-1}(T_l : T_j)\,\rangle = \sum_j k_j \Omega(i,j).$$

The last equality comes from the calculation preceding the definition (5.3d) of the Désarménien coefficient $\Omega(i,j)$. Since (5.6g) lies in $\mathbb{Z}$ for all standard T_i, the unimodularity of the Désarménien matrix shows that $k_j \in \mathbb{Z}$ for all standard T_j. Referring to (5.6e), this proves that $z \in X_{\lambda,\mathbb{Z}}$, and therefore the proof of (5.6d) is complete.

6

Representation theory of the symmetric group

6.1 The functor $\mathbf{f} : \mathbf{M_K(n, r)} \to \mathbf{mod\,KG(r)}$ $(\mathbf{r \leq n})$

In this chapter we shall apply our results on the representations of the general linear group $\Gamma_K = \mathsf{GL}_n(K)$, to the representation theory over K of the symmetric group $G(r)$. The method is to use a process invented by Schur in his dissertation [47]. Suppose first that $r \leq n$. Then there exists a weight $\omega = (1, 1, \ldots, 1, 0, \ldots, 0)$ in $\Lambda(n, r)$ containing r 1's. We shall see that for any module $V \in M_K(n, r)$, the ω-weight-space V^ω can be regarded as a left $KG(r)$-module. The correspondence $V \to V^\omega$ determines a functor $f : M_K(n, r) \to \mathsf{mod}\, KG(r)$. Schur proved (see [47, sections III, IV]) that in case $K = \mathbb{C}$ this functor gives an equivalence between the categories $M_K(n, r)$ and $\mathsf{mod}\, KG(r)$; by this means he showed that modules in $M_{\mathbb{C}}(n, r)$ are completely reducible, hence are determined up to isomorphism by their characters (see [47, p. 35]). The proof which we have given of this fact, see (2.6e), is essentially Schur's later proof in [48, p. 77]. Then Schur was able to handle the case $n < r$ by an argument [47, pp. 61–63] which uses another functor, this time from $M_K(r, r)$ to $M_K(n, r)$. This second functor will be described in 6.5.

Of course Schur used his functor f, and its "inverse" (see 6.2), to make deductions about $M_K(n, r)$—his starting point was the known representation theory of $G(r)$. But since we have already got some knowledge of $M_K(n, r)$ by the "combinatorial" methods of §4, §5, it is also sometimes profitable to work in the other direction.

Let us keep K, n, r fixed for the moment, and write $S = S_K(n, r)$. Any module $V \in M_K(n, r)$ can be regarded as left S-module, and therefore for any weight $\alpha \in \Lambda(n, r)$, the weight-space $V^\alpha = \xi_\alpha V$ (see 3.2) can be regarded as left $S(\alpha)$-module, where $S(\alpha)$ denotes the algebra $\xi_\alpha S \xi_\alpha$. We get then a functor

(6.1a) $\quad f_\alpha : M_K(n, r) \to \mathsf{mod}\, S(\alpha),$

which takes each module $V \in M_K(n,r)$ to $V^\alpha \in \mathsf{mod}\, S(\alpha)$, and each morphism $\theta : V \to V'$ in $M_K(n,r)$ to its restriction $\theta^\alpha : V^\alpha \to (V')^\alpha$.

$S(\alpha)$ is a K-algebra with ξ_α as identity element. If we choose some element $i \in I(n,r)$ which belongs to α, for example

(6.1b) $$i = (\underbrace{1,1,\ldots,1}_{\alpha_1},\underbrace{2,2,\ldots,2}_{\alpha_2},\ldots,\underbrace{n,n,\ldots,n}_{\alpha_n}),$$

we may use the multiplication rules in 2.3 to show that $S(\alpha)$ is spanned, as K-space, by the elements $\xi_{i\pi,i}$, $\pi \in G$. From the equality rule in 3.2 follows that, for any elements $\pi, \pi' \in G$, $\xi_{i\pi,i} = \xi_{i\pi',i}$ if and only if π, π' belong to the same double coset with respect to the subgroup $G_\alpha = \{\pi \in G : i\pi = i\}$ of G. So $S(\alpha)$ has K-basis $\{\xi_{i\pi,i}\}$, π running over a set of representatives of the double-coset space $G_\alpha \backslash G / G_\alpha$.

Now suppose that $r \le n$, and that ω is the weight described above. The element (6.1b) corresponding to $\alpha = \omega$ is written

(6.1c) $$u = (1,2,\ldots,r) \in I(n,r).$$

Since the stabilizer in G of this element is $G_\omega = \{1\}$, the algebra $S(\omega)$ has K-basis $\{\, \xi_{u\pi,u} : \pi \in G \,\}$. An elementary application of multiplication rule (2.3b) shows that $\xi_{u\pi,u}\xi_{u\pi',u} = \xi_{u\pi\pi',u}$, for all π, π' in G. We have therefore an isomorphism of K-algebras

(6.1d) $$S(\omega) \cong KG(r),$$

which takes $\xi_{u\pi,u} \mapsto \pi$ for all $\pi \in G = G(r)$. By means of this isomorphism the categories $\mathsf{mod}\, S(\omega)$ and $\mathsf{mod}\, KG(r)$ can be identified. With this identification we define the *Schur functor* [47, p. 22],

$$f : M_K(n,r) \longrightarrow \mathsf{mod}\, KG(r)$$

to be the functor $f = f_\omega$.

Remark. For the general case, where α is any weight in $\Lambda(n,r)$ (and with no restriction on n, r) $S(\alpha)$ is isomorphic to the *Hecke ring* $\mathsf{H}_K(G, G_\alpha)$ over K. We may follow Iwahori [25, p. 218] and define the Hecke ring $\mathsf{H}(G,H)$ for any subgroup H of any finite group G, as follows. $\mathsf{H}(G,H)$ has free $\mathbb{Z}$-basis $\{\chi_A\}$, where A runs over the set $H\backslash G/H$ of all double-cosets of H in G; the product of elements in this basis is given by

(6.1e) $$\chi_A \chi_B = \sum_{C \in H\backslash G/H} z_{A,B,C}\, \chi_C,$$

where if γ is any fixed element of C, $z_{A,B,C}$ is the number of H-cosets $H\pi$ in the set $A^{-1}\gamma \cap B$. (For an explanation of this artificial-looking rule, see [25, §1].) Alternatively we may define $\mathsf{H}(G,H)$ to be the endomorphism ring of the subset $[H]\mathbb{Z}G$ of $\mathbb{Z}G$, this subset being regarded as right $\mathbb{Z}G$-module. In this interpretation χ_A becomes the $\mathbb{Z}G$-endomorphism of $[H]\mathbb{Z}G$ which takes $[H]$ to $[A]$.

Returning now to our case $G = G(r)$, $H = G_\alpha$, we leave it as an exercise to prove that the K-linear map $S(\alpha) \to \mathsf{H}_K(G, G_\alpha)$ given by $\xi_{i\pi,i} \mapsto \chi_{G_\alpha \pi G_\alpha}$ for all $\pi \in G$, is an isomorphism of K-algebras.

6.2 General theory of the functor f : mod S → mod eSe

It soon becomes clear that many properties of Schur's functor belong to a much more general context. Let S be any K-algebra (it does not need to be finite-dimensional) and let $e \neq 0$ be any idempotent in S. We define a functor $f : \mathsf{mod}\, S \to \mathsf{mod}\, eSe$ as follows. If $V \in \mathsf{mod}\, S$, clearly the subspace eV of V is an eSe-module, so we define $f(V) = eV \in \mathsf{mod}\, eSe$. If $\theta : V \to V'$ is a morphism in $\mathsf{mod}\, S$, then we define $f(\theta) : eV \to eV'$ to be the restriction of θ; clearly $f(\theta)$ is an eSe-morphism. It is important to observe that f *is an exact functor*, in other words

(6.2a) *Suppose* $0 \to V' \to V \to V'' \to 0$ *is an exact sequence in* $\mathsf{mod}\, S$, *then* $0 \to eV' \to eV \to eV'' \to 0$ *is an exact sequence in* $\mathsf{mod}\, eSe$.

This is quite elementary. The next proposition, though easy and undoubtedly well known, does not seem to appear in the literature[1] (a special case is given by Curtis and Fossum [12, p. 402]. Much of the present section 6.2 appears, sometimes with different proofs, in the Ph.D. dissertation of T. Martins [41]).

(6.2b) *If* $V \in \mathsf{mod}\, S$ *is irreducible, then* eV *is either zero or is an irreducible module in* $\mathsf{mod}\, eSe$.

Proof. Let W be any non-zero eSe-submodule of eV. Then $W = eW$, and also $SW = SeW$, which is a non-zero S-submodule of V, is equal to V. Hence $eV = e(SeW) = (eSe)W \subseteq W$. This proves $W = eV$. Therefore if $eV \neq 0$, then eV is an irreducible eSe-module. This proves (6.2b).

Now suppose $V \in \mathsf{mod}\, S$, and define $V_{(e)}$ to be the sum of all the S-submodules V_0 of V such that $eV_0 = 0$—in other words, $V_{(e)}$ is the largest S-submodule of V which is contained in $(1-e)V$. We also define $a(V) = V/V_{(e)}$. Then we can make a functor

$$a : \mathsf{mod}\, S \to \mathsf{mod}\, S;$$

notice that if $\theta : V \to V'$ is a morphism in $\mathsf{mod}\, S$, then θ maps $V_{(e)}$ into $V'_{(e)}$, hence θ induces a well-defined map $a(\theta) : a(V) \to a(V')$. The virtue of this functor, is that it gets rid of the part of each module V which is annihilated by f, and does this without destroying anything in $f(V)$. Expressed precisely, we have

[1]Our functor is a special case of a functor described by M. Auslander in [2]; see p. 243. I am indebted to J. Alperin for this reference.

(6.2c) *Let $V \in \mathsf{mod}\, S$. Then the natural map $\alpha_V : V \to a(V) = V/V_{(e)}$ induces an isomorphism $f(\alpha_V) : f(V) \to f(a(V))$.*

Proof. Clearly $f(\alpha_V)$, which is just the restriction of α_V to $f(V) = eV$, is *onto* $f(a(V)) = e \cdot a(V)$. And $\mathsf{Ker}\, f(\alpha_V) = eV \cap V_{(e)} = 0$ since $V_{(e)} \subseteq (1-e)V$. Thus $f(\alpha_V)$ is an isomorphism.

Our next objective is to define functors from $\mathsf{mod}\, eSe$ to $\mathsf{mod}\, S$, which can serve, at least partially, as inverses to f. As first attempt we employ the definition

$$h(W) = Se \otimes_{eSe} W, \qquad \text{for } W \in \mathsf{mod}\, eSe.$$

Since Se is a left S-module (it is a left ideal of S, of course) and also a right eSe-module, $h(W)$ is well-defined and is a left S-module. If $\psi : W \to W'$ is a morphism in $\mathsf{mod}\, eSe$, then $h(\psi) = \mathbf{1}_{Se} \otimes \psi : h(W) \to h(W')$ is a morphism in $\mathsf{mod}\, S$. We get in this way a functor $h : \mathsf{mod}\, eSe \to \mathsf{mod}\, S$. Moreover the next proposition shows that h is a "right-inverse" to f.

(6.2d) *Let $W \in \mathsf{mod}\, eSe$. Then $e \cdot h(W) = e \otimes W$, and the map $w \mapsto e \otimes w$ ($w \in W$) gives an eSe-isomorphism $W \cong e \cdot h(W) = f(h(W))$.*

Proof. $e \cdot h(W) = e(Se \otimes_{eSe} W) = eSe \otimes_{eSe} W = e \otimes W$, as stated. Thus the map defined above takes W onto $e \cdot h(W)$; it is elementary to check that it is an eSe-map. To prove that it is injective, first notice that there is a well-defined map $\eta : Se \otimes_{eSe} W \to W$ such that $\eta(s \otimes w) = esw$, for all $s \in Se$, $w \in W$. Then if $w \in W$ is such that $e \otimes w = 0$, we get $0 = \eta(e \otimes w) = w$. This establishes the injectivity of the map $w \mapsto e \otimes w$, and (6.2d) is proved.

The trouble with the functor h is that it usually takes an irreducible module W to a module $h(W)$ which is not irreducible. However we have

(6.2e) *If $W \in \mathsf{mod}\, eSe$ is irreducible, then $h(W)_{(e)}$ is the unique maximal proper submodule of $h(W)$. Hence $a(h(W))$ is irreducible.*

Proof. Write $V = h(W)$. Then by (6.2d) and (6.2c), $f(a(V)) \cong f(V) \cong W$. Thus $a(V) \neq 0$, which shows that $V_{(e)}$ is a proper submodule of V. Now let V' be any proper submodule of V. If $eV' \neq 0$ then eV', being an eSe-submodule of the irreducible eSe-module $V = e \cdot h(W)$ (recall $e \cdot h(W) = e \otimes W \cong W$, by (6.2d)) is equal to $e \cdot h(W)$. Then $V' \supseteq SeV' = S(e \otimes W) = h(W) = V$, a contradiction. So $eV' = 0$, i.e. V' is contained in $V_{(e)}$. This proves (6.2e).

Definition. Let h^* denote the functor $ah : \mathsf{mod}\, eSe \to \mathsf{mod}\, S$, so that

$$h^*(W) = h(W)/h(W)_{(e)}$$

for all $W \in \mathsf{mod}\, eSe$.

By (6.2c) and (6.2d) this functor h^*, like h, is a right inverse to f, i.e. $f(h^*(W)) \cong W$ for all $W \in \mathsf{mod}\, eSe$. By (6.2e) h^* takes irreducibles to irreducibles. We have finally

(6.2f) *If $V \in \mathsf{mod}\, S$ is irreducible and if $eV \neq 0$, then $h^*(eV) \cong V$.*

Proof. There is an S-map $\beta : h(eV) = Se \otimes_{eSe} eV \to V$, which takes $s \otimes ev$ to sev, for all $s \in Se$, $v \in V$. The image of β is SeV, which equals V because V is irreducible. So the kernel of S is a maximal proper submodule of $h(eV)$. But $eV \in \mathsf{mod}\, eSe$ is irreducible by (6.2b), hence the only maximal proper submodule of $h(eV)$ is $h(eV)_{(e)}$, by (6.2e). Therefore β induces an isomorphism of $h(eV)/h(eV)_{(e)} = a(h(eV)) = h^*(eV)$ onto V.

Taking together all these facts, we arrive at our main theorem.

(6.2g) Theorem. *Suppose $\{V_\lambda : \lambda \in \Lambda\}$ is a full set of irreducible modules in $\mathsf{mod}\, S$, indexed by a set Λ, and let $\Lambda' = \{\lambda \in \Lambda : eV_\lambda \neq 0\}$. Then $\{eV_\lambda : \lambda \in \Lambda'\}$ is a full set of irreducible modules in $\mathsf{mod}\, eSe$. Moreover if $\lambda \in \Lambda'$, then $V_\lambda \cong h^*(eV_\lambda)$.*

Remarks. 1. It is well-known that $\mathsf{Hom}_S(Se, V) \cong eV$ (as K-spaces), for any $V \in \mathsf{mod}\, S$ (see [11, p. 375]). Therefore if V is irreducible, $eV \neq 0$ if and only if V is a homomorphic image of Se.

2. When we come to apply the Schur functor to the Carter-Lusztig modules $V_{\lambda,K}$, it will be useful to notice that if any $V \in \mathsf{mod}\, S$ has a unique maximal proper submodule $V^{\max}$, then $eV^{\max}$ is either equal to eV (i.e. $e(V/V^{\max}) = 0$) or else it is the unique maximal proper submodule of the eSe-module eV. The proof is easy.

3. In the same context we shall use the following: If $(\ ,\ )$ is a symmetric bilinear form on V such that $(eV, (1-e)V) = 0$, and if $(\ ,\ )^e$ denotes the restriction of this form to eV, then $\mathsf{rad}\,(\ ,\ )^e = e \cdot \mathsf{rad}(\ ,\ )$. Again, the proof is an easy exercise.

6.3 Application I. Specht modules and their duals

In this section we shall apply the general theory of 6.2 to the special case of the Schur functor $f : M_K(n,r) \to \mathsf{mod}\, KG(r)$. Here K is any infinite field, and n, r are fixed integers *such that* $r \leq n$. We take $S = S_K(n,r)$ and $e = \xi_\omega = \xi_{u,u}$ (see 6.1 for notation), and identify eSe with $KG(r)$ by the isomorphism (6.1d), which takes $\xi_{u\pi,u} \mapsto \pi$, for all $\pi \in G = G(r)$. Notice that $f(V) = eV = V^\omega$, for any $V \in M_K(n,r)$.

Our aim is to calculate the effect of f on the modules $D_{\lambda,K}$, $V_{\lambda,K}$. Notice that the elements λ of $\Lambda^+(n,r)$ are in one-to-one correspondence with the partitions λ of r (because $r \leq n$). We shall write $\Lambda = \Lambda^+(n,r)$, and think of Λ as the set of all partitions of r. From now on, λ is a fixed element of Λ.

Recall (p. 35) that $D_{\lambda,K}$ is the K-span of the elements

$$(T_l : T_i) = \sum_{\sigma \in C(T)} \mathsf{s}(\sigma)\, c_{l,i\sigma}\,, \qquad \text{all } i \in I(n,r).$$

We saw (p. 37) that these bideterminants $(T_l : T_i)$ all lie in the right λ-weight-space ${}^\lambda A_K(n,r) = A_K(n,r) \circ \xi_\lambda$ of $A_K(n,r)$. But by definition $f(D_{\lambda,K})$ is the ω-weight-space $D^\omega_{\lambda,K}$ of $D_{\lambda,K}$, and therefore lies in the (left) ω-weight-space

(6.3a) ${}^\lambda A_K(n,r)^\omega = \xi_\omega \circ A_K(n,r) \circ \xi_\lambda$

of ${}^\lambda A_K(n,r)$. Elementary calculations based on formulae (4.4a), (4.4a′), (6.3a) show that ${}^\lambda A_K(n,r)^\omega = \sum_{\pi \in G} K \cdot c_{l,u\pi}$. Since $c_{l,u\pi} = c_{l,u\pi'}$ if and only if $\pi' \in \pi R(T)$ (see 4.5), there is an isomorphism of K-spaces

(6.3b) ${}^\lambda A_K(n,r)^\omega \to KG[R(T)]$

which takes $c_{l,u\pi} \mapsto \pi[R(T)]$, for all $\pi \in G$. Now $KG[R(T)]$ is a left ideal of the group algebra KG, hence is a left KG-module. On the other hand ${}^\lambda A_K(n,r)^\omega$, being the ω-weight-space of the left $S_K(n,r)$-module ${}^\lambda A_K(n,r)$, becomes a left KG-module by means of (6.1d). To be explicit, the element $\tau \in G$ acts on the element $c_{l,u\pi}$ to give $\tau c_{l,u\pi} = \xi_{u\tau,u} \circ c_{l,u\pi}$, which by (4.4a) is equal to $c_{l,u\tau\pi}$. It follows at once that *(6.3b) is a left KG-isomorphism.*

By (4.7a), p. 39, $f(D_{\lambda,K}) = D^\omega_{\lambda,K}$ has K-basis consisting of all $(T_l : T_i)$ such that $i \in \omega$ and T_i is standard. The elements i in ω can be written, uniquely, in the form $i = u\pi$ $(\pi \in G)$. The isomorphism (6.3b) takes $(T_l : T_{u\pi})$ to

$$\sum_{\sigma \in C(T)} \mathsf{s}(\sigma)\, \pi\sigma\, [R(T)] = \pi\, \{C(T)\}\, [R(T)],$$

and so it takes $f(D_{\lambda,K})$ to the left KG-submodule (left ideal)

$$S_{T,K} = KG\, \{C(T)\}\, [R(T)]$$

of KG. We shall define $S_{T,K}$ to be the *Specht module* (over K) corresponding to the bijective λ-tableau T. (This is a little different from the original definition of Specht; for an explanation of the latter, and of the equivalence of the two definitions, see [45, p. 91].) We have now the

(6.3c) Theorem. *The Specht module $S_{T,K}$ has K-basis consisting of the elements $\pi\, \{C(T)\}\, [R(T)]$ such that $T_{u\pi}$ is standard. If* $\mathsf{char}\, K = 0$*, then $S_{T,K}$ is an irreducible KG-module. If we choose for every $\lambda \in \Lambda$ a bijective λ-tableau T^λ, and write $S_{\lambda,K} = S_{T^\lambda,K}$, then $\{S_{\lambda,K} : \lambda \in \Lambda\}$ is a full set of irreducible KG-modules.*

Proof. The first statement comes by applying the isomorphism (6.3b) to the basis of $D_{\lambda,K}$ given by (4.7a), already quoted.

If $\mathsf{char}\, K = 0$, then each $D_{\lambda,K}$ is irreducible by (4.7b), and since $D_{\lambda,K}$ has character S_λ (by (4.7a)), $\{ D_{\lambda,K} : \lambda \in \Lambda\}$ is a full set of irreducible modules in $M_K(n,r)$. Then the last statement in (6.3c) follows at once from (6.2g), since in this present case $f(D_{\lambda,K}) \cong S_{\lambda,K}$ is non-zero for all $\lambda \in \Lambda$.

Now let's look at the module $V_{\lambda,K}$. By definition (see (5.1b)) this is a subspace of $E^{\otimes r}$, therefore $f(V_{\lambda,K}) = V^{\omega}_{\lambda,K}$ is a subspace of $f(E^{\otimes r}) = (E^{\otimes r})^{\omega}$. From the formula (2.6a) which gives the action of $S = S_K(n,r)$ on $E^{\otimes r}$, we see that for any weight $\alpha \in \Lambda(n,r)$ (and with no restriction on n, r)

$$(E^{\otimes r})^{\alpha} = \xi_{\alpha}\, E^{\otimes r} = \sum_{i \in \alpha} K \cdot e_i \,.$$

So in particular $(E^{\otimes r})^{\omega}$ has K-basis $\{\, e_{u\pi} : \pi \in G \,\}$. The structure of $(E^{\otimes r})^{\omega}$ as left KG-module is given by $\tau e_{u\pi} = \xi_{u\tau,u} e_{u\pi} = e_{u\tau\pi}$, for all $\tau, \pi \in G$. Therefore there is a left KG-module isomorphism

(6.3d) $\quad (E^{\otimes r})^{\omega} \to KG,$

which takes $e_{u\pi} \mapsto \pi$, for all $\pi \in G$.

We know that $V_{\lambda,K}$ has K-basis $\{\, b_{u\pi} : \pi \in G,\ T_{u\pi} \text{ standard} \,\}$, by (5.3b) and (5.4a). Recall that, for any $i \in I(n,r)$, $b_i = \xi_{i,l} f_l = \xi_{i,l} e_l \{C(T)\}$ (see 5.3). If we put $i = u\pi$ in formula (5.3c) we get $\xi_{u\pi,l} e_l = e_u \pi [R(T)]$. This means $b_{u\pi} = e_u \pi [R(T)]\{C(T)\}$. This element is carried by the isomorphism (6.3d) to the element $\pi[R(T)]\{C(T)\}$ of KG. Therefore $V_{\lambda,K}$ is carried to the left KG-submodule (left ideal)

$$\overline{S}_{T,K} = KG\,[R(T)]\,\{C(T)\}$$

of KG. We have the following theorem, whose proof is entirely analogous to that of (6.3c).

(6.3e) Theorem. *The module $\overline{S}_{T,K}$ defined above has K-basis consisting of the elements $\pi\,[R(T)]\,\{C(T)\}$ such that $T_{u\pi}$ is standard. If* char $K = 0$, *then $\overline{S}_{T,K}$ is an irreducible KG-module. If we choose for each $\lambda \in \Lambda$ a bijective λ-tableau T^{λ}, and write $\overline{S}_{\lambda,K} = \overline{S}_{T^{\lambda},K}$, then $\{\overline{S}_{\lambda,K} : \lambda \in \Lambda\}$ is a full set of irreducible KG-modules.*

The module $\overline{S}_{T,K}$ is in fact the dual (in the usual sense) to the Specht module $S_{T,K}$—this was first proved by G. D. James [26, p. 460]. We can give another proof: the modules $V_{\lambda,K}$, $D_{\lambda,K}$ are dual to each other under the contravariant form (,) described in 5.1. Therefore $V^{\omega}_{\lambda,K}$, $D^{\omega}_{\lambda,K}$ are dual to each other under the restriction of this form (see 3.3). The contravariant property (2.7d) gives

$$(\xi_{u\pi,u} v, d) = (v, \xi_{u,u\pi} d) = (v, \xi_{u\pi^{-1},u} d),$$

for all $\pi \in G$, $v \in V^{\omega}_{\lambda,K}$, $d \in D^{\omega}_{\lambda,K}$. But this becomes $(\pi v, d) = (v, \pi^{-1} d)$ when we regard $V^{\omega}_{\lambda,K}$, $D^{\omega}_{\lambda,K}$ as KG-modules by means of (6.1d), and this shows that these KG-modules are dual to each other. Naturally we can transfer this form, by means of the isomorphisms (6.3b), (6.3d), to give an invariant form $\overline{S}_{T,K} \times S_{T,K} \to K$. We leave it to the reader to do this, and also to apply the calculation given in 5.3 to exhibit the following explicit version of the invariant form in question.

(6.3f) Theorem. *The KG-modules $\overline{S}_{T,K}$, $S_{T,K}$ are dual to each other. There is an invariant bilinear form $(\ ,\ ) : \overline{S}_{T,K} \times S_{T,K} \to K$ such that*

$$\Big(\pi\,[R(T)]\,\{C(T)\}, \pi'\{C(T)\}\,[R(T)]\Big) = \Omega_{\pi,\pi'}$$

for all $\pi, \pi' \in G$, where $\Omega_{\pi,\pi'} = \sum \mathsf{s}(\sigma)$, sum over all $\sigma \in C(T)$ such that the tableaux πT and $\pi'\sigma T$ are row-equivalent.

The matrix $(\,\Omega_{\pi,\pi'}\,:\,\pi T,\,\pi' T$ standard$)$ is unipotent triangular, relative to a suitable ordering of the standard πT.

Remarks. 1. Since $u = (1, 2, \ldots, r)$, the tableau T_u can be identified with the basic tableau T, and $T_{u\pi}$ with πT.

2. The matrix $(\Omega_{\pi,\pi'})$ appearing in (6.3f) is just that part of the Désarménien matrix $(\Omega(i,j))$ corresponding to $i, j \in \omega$; in fact $\Omega_{\pi,\pi'} = \Omega(u\pi, u\pi')$.
3. All the results in this section remain true when K is replaced by $\mathbb{Z}$. For if we take $K = \mathbb{Q}$, then we may check that the isomorphisms (6.3b), (6.3d) take $D^{\omega}_{\lambda,\mathbb{Z}}$, $V^{\omega}_{\lambda,\mathbb{Z}}$ to $S_{T,\mathbb{Z}} = \mathbb{Z}G\,\{C(T)\}\,[R(T)]$, $\overline{S}_{T,\mathbb{Z}} = \mathbb{Z}G\,[R(T)]\,\{C(T)\}$, respectively. Therefore these last are $\mathbb{Z}$-forms of the $\mathbb{Q}G$-modules $S_{T,\mathbb{Q}}$ and $\overline{S}_{T,\mathbb{Q}}$, respectively, with $\mathbb{Z}$-bases $\{\pi\,\{C(T)\}\,[R(T)]\,:\,T_{u\pi}$ standard$\}$ and $\{\pi\,[R(T)]\,\{C(T)\}\,:\,T_{u\pi}$ standard$\}$.

6.4 Application II. Irreducible KG(r)-modules, char K = p

Throughout this section we assume that K has finite characteristic p, and that r, n are positive integers satisfying $r \leq n$. In 5.4 we constructed a full set $\{\,F_{\lambda,K}\,:\,\lambda \in \Lambda^+(n,r) = \Lambda\,\}$ of irreducible modules in $M_K(n,r)$. Apply the Schur functor f, and we have by (6.2g) the theorem

(6.4a) *Let Λ be the set of all partitions of r, and let Λ' be the subset of Λ consisting of those λ such that $F^{\omega}_{\lambda,K} \neq 0$. Then $\{\,F^{\omega}_{\lambda,K}\,:\,\lambda \in \Lambda'\,\}$ is a full set of irreducible $KG(r)$-modules.*

Of course this still leaves open the crucial question: what is the set Λ'? The answer is contained in the next theorem.

(6.4b) Theorem (Clausen[2], James[3]). *The set Λ' of (6.4a) consists of those partitions $\lambda = (\lambda_1, \lambda_2, \ldots, \lambda_r, 0, \ldots)$ of r which are "column p-regular" i.e. for which all the integers $\lambda_1 - \lambda_2, \lambda_2 - \lambda_3, \ldots, \lambda_r$ lie between 0 and $p-1$.*

---

[2] [8, Lemma 6.4, p. 184]
[3] [28, Theorem 3.2]

Proof. It will be convenient to work, not with $F_{\lambda,K}$, but with the isomorphic module $D^{\min}_{\lambda,K}$ (see (5.4c)). We denote this module by X. We must show that $X^\omega \neq 0$ if and only if λ is column p-regular.

By (5.4d), X is generated as $S_K(n,r)$-module by $(T_l : T_l)$. Therefore it is spanned as K-space by the elements $\xi_{i,j} \circ (T_l : T_l)$, for all $i, j \in I = I(n,r)$. By (4.4b)

$$\xi_{i,j} \circ (T_l : T_l) = \sum_{h \in I} \xi_{i,j}(c_{h,l})\,(T_l : T_h),$$

which is zero unless $j \sim l$. Therefore X is K-spanned by the elements

(6.4c) $$\xi_{i,l} \circ (T_l : T_l) = \sum_{h \in I} \xi_{i,l}(c_{h,l})\,(T_l : T_h) = \sum_{h \in iR(T)} (T_l : T_h), \text{ all } i \in I.$$

The element (6.4c) lies in the α-weight-space X^α, where α is the weight containing i. So X^ω is K-spanned by those elements (6.4c) such that $i \in \omega$. If $i \in \omega$, then G acts regularly on $iG = \omega$ i.e. $i\pi = i\pi'$ implies $\pi = \pi'$, for all $\pi, \pi' \in G$. In particular the elements $i\tau$ ($\tau \in R(T)$) are all distinct. So

(6.4d) If $i \in \omega$, then $\xi_{i,l} \circ (T_l : T_l) = \sum_{\tau \in R(T)} (T_l : T_{i\tau})$.

Suppose that H is the group of all elements θ of $R(T)$ which preserve the set of columns of the basic λ-tableau T. Such a permutation θ can be specified by a sequence $\theta_1, \theta_2, \ldots$, where for each $q \geq 1$, θ_q is a permutation of the set W_q of all $t \geq 1$ such that column t of T has length q. In the notation of (4.2a), θ maps $x(s,t)$ to $x(s, \theta_q(t))$, for all $s \geq 1$, and all $t \in W_q$. Since $|W_q| = \lambda_q - \lambda_{q+1}$, the order of H is $(\lambda_1 - \lambda_2)!\,(\lambda_2 - \lambda_3)! \cdots$. Now it follows from the expression of $(T_l : T_i)$ as product of determinants (see 4.3) that $(T_l : T_i) = (T_l : T_{i\theta})$, for all $i \in I$ and all $\theta \in H$. So by breaking up the sum in (6.4d) into H-orbits, we see that it is divisible by $|H|$. If λ is column p-singular, $|H|$ is divisible by p, hence every term (6.4d) is zero, i.e. $X^\omega = 0$.

This proves one half of (6.4b). To prove the other half, we assume that λ is column p-regular, and show that $X^\omega \neq 0$. For this it is enough to show that $\xi_{u,l} \circ (T_l : T_l) \neq 0$. By (6.4d) and (4.3a), $\xi_{u,l} \circ (T_l : T_l)$ is equal to

(6.4e) $$\sum_{\sigma \in C(T)} \sum_{\tau \in R(T)} \mathsf{s}(\sigma)\, c_{l\sigma, u\tau}.$$

There is a unique element $\pi \in C(T)$ which reverses the order of the entries in each column of T_l, namely

$$\pi : x(s,t) \mapsto x(q+1-s, t), \qquad \text{for } s \geq 1, \text{ and } t \in W_q.$$

For example if $\lambda = (7,5,2,2)$ we have

$$T_l = \begin{matrix} 1\,1\,1\,1\,1\,1\,1 \\ 2\,2\,2\,2\,2 \\ 3\,3 \\ 4\,4 \end{matrix}, \qquad T_{l\pi} = \begin{matrix} 4\,4\,2\,2\,2\,1\,1 \\ 3\,3\,1\,1\,1 \\ 2\,2 \\ 1\,1 \end{matrix}.$$

We shall prove that the coefficient of $c_{l\pi,u}$ in (6.4e) is not zero. If $\sigma \in C(T)$ and $\tau \in R(T)$ are such that $c_{l\pi,u} = c_{l\sigma,u\tau}$, then there is some $\gamma \in G$ such that $l\pi\gamma = l\sigma$, $u\gamma = u\tau$. This implies $\gamma = \tau$, hence $l\pi\tau = l\sigma$. Consider the maximum entry in T_l, say M (in the example, $M = 4$). In T_l, and hence also in $T_{l\sigma}$, all entries M are in the columns $t \in W_M$. But in $T_{l\pi}$, hence also in $T_{l\pi\tau}$, all entries M are in the first row. Since $T_{l\pi\tau} = T_{l\sigma}$, all entries M in $T_{l\sigma}$ are in the same places as the entries M in $T_{l\pi}$. Next we consider the places occupied by entries $M-1, M-2, \ldots$ in turn, and by arguments similar to that given conclude that $T_{l\pi} = T_{l\sigma}$, hence that $\pi = \sigma$. The group of elements $\tau \in R(T)$ satisfying $l\pi\tau = l\pi$ clearly has order

$$w = \prod_{q \geq 1} \left((\lambda_q - \lambda_{q+1})! \right)^q,$$

which is prime to p. The argument just given shows that the coefficient of $c_{l\pi,u}$ in (6.4e) is $\mathsf{s}(\pi)\, w \cdot 1_K \neq 0$, and so the proof of (6.4b) is complete.

This theorem has some interesting consequences. First we need a lemma concerning the left ideal $S\xi_\omega$ of S, which is also a right module for the algebra $S(\omega) = \xi_\omega S \xi_\omega$, and hence, by (6.1d), a right KG-module.

(6.4f) *$S\xi_\omega$ has K-basis $\{ \xi_{i,u} : i \in I(n,r) \}$. The K-isomorphism $S\xi_\omega \to E^{\otimes r}$ given by $\xi_{i,u} \mapsto e_i$ for all $i \in I(n,r)$ is a left $S = S_K(n,r)$-map and a right KG-map.*

The proof of (6.4f) is routine. Now let $V \in M_K(n,r)$ be irreducible. By 6.2, remark 1, $V^\omega \neq 0$ if and only if V is a homomorphic image of $S\xi_\omega$, and hence of $E^{\otimes r}$. But both $E^{\otimes r}$ and V are self-dual (by 2.7, Example 1, and (5.4c), proof). We have therefore

(6.4g) *If $V \in M_K(n,r)$ is irreducible, then $V^\omega \neq 0$ if and only if V is isomorphic to a submodule of $E^{\otimes r}$. (Notice, we assume $r \leq n$.)*

Corollary (James [28, Theorem 3.2]). *$F_{\lambda,K}$ is isomorphic to a submodule of $E^{\otimes r}$ if and only if λ is column p-regular.*

Next we have a theorem concerning the "dual" Specht module

$$\overline{S}_{T,K} = KG\,[R(T)]\,\{C(T)\}$$

of 6.3. In 5.5 was defined a contravariant form $\langle\langle\ ,\ \rangle\rangle$ on the space $E^{\otimes r}\{C(T)\}$. Restrict this to the ω-weight-space $(E^{\otimes r})^\omega\{C(T)\}$, and then transfer it to $KG\{C(T)\}$ by means of the isomorphism (6.3d). The result is a symmetric, invariant form on $KG\{C(T)\}$ which we denote by $(\ ,\ )$, and which is specified by the formula: if $\pi, \pi' \in G$, then

$$\textbf{(6.4h)} \qquad \left(\pi\,\{C(T)\}, \pi'\,\{C(T)\} \right) = \begin{cases} \mathsf{s}(\pi^{-1}\pi') & \text{if } \pi^{-1}\pi' \in C(T), \\ 0 & \text{if } \pi^{-1}\pi' \notin C(T). \end{cases}$$

In 5.5 we considered the form obtained by restricting $\langle\langle\ ,\ \rangle\rangle$ to $V_{\lambda,K}$, and showed (see (5.5c)) that the radical of this form is the unique maximal submodule $V_{\lambda,K}^{\max}$ of $V_{\lambda,K}$. It is a routine matter now to apply the Schur functor, and use remarks 2, 3 of 6.2 to prove the following.

(6.4i) *Let* $(\ ,\ )$ *be the invariant form on* $\overline{S}_{T,K}$ *obtained by restricting the form given by (6.4h). Then* $(\ ,\ )$ *is non-zero on* $\overline{S}_{T,K}$ *if and only if* λ *is column p-regular. If* λ *is column p-regular, then the radical of* $(\ ,\ )$ *is the unique maximal submodule* $\overline{S}_{T,K}^{max}$ *of* $\overline{S}_{T,K}$, *and* $\overline{S}_{T,K}/\overline{S}_{T,K}^{max} \cong f(F_{\lambda,K})$.

From (6.4i) we may deduce a well-known theorem of James (see (6.4k), below), by the following elementary device. Let β denote the K-algebra automorphism of KG given by $\beta(\pi) = \mathsf{s}(\pi)\,\pi$, for all $\pi \in G$. Let K_{s} denote the field K, regarded as one-dimensional KG-module by the action $\pi k = \mathsf{s}(\pi)\,k$, for $\pi \in G$, $k \in K$. Then if M is any left ideal of KG, $\beta(M)$ is also a left ideal of KG, and there is a KG-isomorphism $\beta(M) \cong M \otimes_K K_{\mathsf{s}}$ which takes $m \otimes 1_K \mapsto \beta(m)$, for all $m \in M$. It is trivial to check that β maps $\{C(T)\}$, $[R(T)]$ to $[R(T')]$, $\{C(T')\}$ respectively, where T' is the λ'-tableau (λ' is the partition of r conjugate to λ) obtained by "transposing" the λ-tableau T. The bilinear form (6.4h) on $KG\{C(T)\}$ is translated by β to a symmetric, invariant bilinear form $(\ ,\ )$ on $KG[R(T')]$ specified by the formula: if $\pi, \pi' \in G$, then

$$\textbf{(6.4j)}\qquad \Big(\pi\,[R(T')], \pi'\,[R(T')]\Big) = \begin{cases} 1 \text{ if } \pi^{-1}\pi' \in R(T'), \\ 0 \text{ if } \pi^{-1}\pi' \notin R(T'). \end{cases}$$

Moreover $\beta(\overline{S}_{T,K}) = S_{T',K}$—which shows incidentally that

$$S_{T',K} \cong \overline{S}_{T,K} \otimes_K K_{\mathsf{s}}$$

—and so (6.4i) translates as follows.

(6.4k) Theorem (James [27, Theorems 11.1, 11.5]). *Let* T' *be a bijective* μ*-tableau, where* μ *is a partition of* r. *Let* $(\ ,\ )$ *be the invariant form on* $S_{T',K}$ *obtained by restricting the form given by (6.4j). Then* $(\ ,\ )$ *is non-zero on* $S_{T',K}$ *if and only if* μ *is p-regular (by definition,* μ *is p-regular if* μ' *is column p-regular). If* μ *is p-regular, then the radical of* $(\ ,\ )$ *is the unique maximal submodule* $S_{T',K}^{max}$ *of* $S_{T',K}$, *and* $S_{T',K}/S_{T',K}^{max} \cong f(F_{\mu',K}) \otimes_K K_{\mathsf{s}}$.

Remark. Comparison with the notation of James in [27], shows that the module D^{μ} [27, p. 39] is isomorphic to $f(F_{\mu',K}) \otimes_K K_{\mathsf{s}}$. The module D_{λ} in [27, §1] is isomorphic to $f(F_{\lambda,K})$. So the connection between James's two families of irreducible KG-modules is

(6.4l) $D^{\lambda'} \cong D_{\lambda} \otimes_K K_{\mathsf{s}}$, for all column p-regular λ.

The importance of James's theorem, or of the equivalent theorem (6.4i), is that it gives a satisfactory "natural" labelling of the (isomorphism classes of) irreducible KG-modules: for each column p-regular λ, D_λ is isomorphic to the unique irreducible quotient of a dual Specht module $\overline{S}_{\lambda,K}$. We have also seen that Schur's functor f gives an independent connection, $D_\lambda \cong f(V_{\lambda,K})$. The Schur functor has a one-sided inverse, namely the functor h^* defined in 6.2. It is interesting that h^* is related to a construction used by James in [28].

First we re-define h, h^* as functors from $\mathsf{mod}\, KG(r) \to M_K(n,r)$, using the isomorphism $S\xi_\omega \cong E^{\otimes r}$ of (6.4f), and the isomorphism $\xi_\omega S \xi_\omega \cong KG(r)$ of (6.1d). This means that for each $W \in \mathsf{mod}\, KG(r)$ we define

$$h(W) = E^{\otimes r} \otimes_{KG} W, \qquad h^*(W) = h(W)/h(W)_{(\omega)},$$

where for any $V \in M_K(n,r)$, $V_{(\omega)}$ is the sum of all S-submodules U of V such that $U^\omega = 0$. Since the isomorphism (6.4f) takes ξ_ω to e_u, it follows from (6.2d) that $h(W)^\omega = e_u \otimes W$.

(6.4m) *Let W be any left ideal of KG, regarded as left KG-module. Define the S-map $\gamma : h(W) \to E^{\otimes r} W$ by $\gamma(x \otimes w) = xw$, for all $x \in E^{\otimes r}$, $w \in W$. Then $\mathsf{Ker}\,\gamma = h(W)_{(\omega)}$. Hence γ induces an S-isomorphism $h^*(W) \cong E^{\otimes r} W$.*

Proof. Suppose that $V = \mathsf{Ker}\,\gamma$. Any element v of V^ω can be written $v = e_u \otimes w$ for some $w \in W$, since $V^\omega \subseteq h(W)^\omega = e_u \otimes W$. But we have $0 = \gamma(v) = e_u w$. This implies $w = 0$, since the elements $e_u \pi = e_{u\pi}$ $(\pi \in G)$ form a K-basis of $e_u KG$. Hence $v = 0$, i.e. $V \subseteq h(W)_{(\omega)}$.

If $\gamma(h(W)_{(\omega)})$ is not zero, it contains some irreducible submodule M. Since the ω-weight-space of $h(W)_{(\omega)}$ is zero, the same must be true for M. But M, as irreducible submodule of $E^{\otimes r}$, satisfies $M^\omega \neq 0$ by (6.4g). This contradiction implies $h(W)_{(\omega)} \subseteq V$, and the rest of the proof of (6.4m) is immediate.

Since KG is a Frobenius algebra [11, p. 420], every irreducible left KG-module is isomorphic to some left ideal W of KG [11, p.417, (61.6)]. If we combine this with (6.4m) and (6.2g), we have a way of constructing some irreducible $S_K(n,r)$-modules (this method is due to James [28]). Of course we can get in this way only modules isomorphic to $F_{\lambda,K}$ for λ column p-regular.

Example. $W = K[G]$ is an ideal of KG, and as left KG-module W affords the trivial representation of G. By (6.4m), $h^*(W)$ is isomorphic to the S-module $E^{\otimes r} W = E^{\otimes r}[G]$. If $i \in \alpha \in \Lambda(n,r)$, then $e_i[G] = |G_\alpha|\, v_\alpha$, where v_α is the basis element of $V_{r,K}$ described in 5.2, Example 1. So $E^{\otimes r} W$ is a submodule of $V_{r,K}$, and has weights α, for all α such that p does not divide $|G_\alpha| = \alpha_1! \cdots \alpha_n!$. So $h^*(W) \cong F_{\lambda,K}$, where λ is the highest such weight. It is easy to see that $\lambda = (p-1, \ldots, p-1, s, 0, \ldots)$, where there are q terms $p-1$, and the non-negative integers q, s are given by $r = q(p-1) + s$.

6.5 Application III. The functor $\mathbf{f : M_K(N,r) \to M_K(n,r)}$ $(\mathbf{N \geq n})$

In this section we fix our infinite field K, and also the integer $r \geq 0$. We consider connections between categories $M_K(n,r)$, as n varies. Suppose that N, n are positive integers such that $N \geq n$. We shall produce a functor

$$d = d_{N,n} : M_K(N,r) \to M_K(n,r),$$

which can be viewed as special case of our general "mod $S \to$ mod eSe" functor of 6.2. The functor d gives a very easy way of passing from the irreducible modules in $M_K(N,r)$ to those in $M_K(n,r)$, and behaves in a very satisfactory way with regard to characters (see (6.5b)). It is based on a construction given by Schur in his dissertation [47, p. 61].

Since $N \geq n$, we may regard $I(n,r)$ as a subset of $I(N,r)$, namely $I(N,r)$ consists of all $i = (i_1, \ldots, i_r)$ with components $i_\varrho \in \underline{N}$, and $I(n,r)$ is the subset of those i whose components all lie in $\underline{n}$. With this convention, $S_K(n,r)$ can be regarded as a subalgebra of $S_K(N,r)$. For $S_K(N,r)$ has basis

(6.5a) $\{ \xi_{i,j} : i,j \in I(N,r) \}$,

and we can identify $S_K(n,r)$ with the K-subspace spanned by those $\xi_{i,j}$ for which $i,j \in I(n,r)$. For the rule (2.3b) for multiplying two elements $\xi_{i,j}$, $\xi_{k,l}$ of type (6.5a), has the consequence that if $i,j,k,l \in I(n,r)$, then the coefficient of any $\xi_{p,q}$ in the product $\xi_{i,j}\xi_{k,l}$ is zero unless both $p,q \in I(n,r)$, while if p, q do belong to $I(n,r)$, then this coefficient is the same as it would be if $\xi_{i,j}\xi_{k,l}$ were computed in $S_K(n,r)$.

We define an injective map $\alpha \mapsto \alpha^*$ of $\Lambda(n,r)$ into $\Lambda(N,r)$ as follows. If $\alpha = (\alpha_1, \ldots, \alpha_n) \in \Lambda(n,r)$, we define $\alpha^* \in \Lambda(N,r)$ by

$$\alpha^* = (\alpha_1, \ldots, \alpha_n, 0, \ldots, 0).$$

Then the image of $\Lambda(n,r)$ under this map is the set

$$\Lambda(n,r)^* = \{ \beta \in \Lambda(N,r) : \beta_{n+1} = \cdots = \beta_N = 0 \}.$$

Notice that if i belongs to a weight $\beta \in \Lambda(n,r)^*$, then $i \in I(n,r)$. So another description of $\Lambda(n,r)^*$ is that it is the set of those $G(r)$-orbits of $I(N,r)$ which lie in $I(n,r)$.

Now define the following element of $S_K(N,r)$:

(6.5b) $e = \sum_\beta \xi_\beta$, sum over all $\beta \in \Lambda(n,r)^*$.

This is an idempotent of $S_K(N,r)$, and it is clear (using (2.3c)) that $e\xi_{i,j} = \xi_{i,j}$ or zero, according as $i \in I(n,r)$ or not, and $\xi_{i,j}e = \xi_{i,j}$ or zero, according as $j \in I(n,r)$ or not. It follows at once that

$$eS_K(N,r)e = S_K(n,r).$$

From 6.2 (taking $S = S_K(N,r)$) we have a functor $d : M_K(N,r) \to M_K(n,r)$ which takes each $V \in M_K(N,r)$ to $eV \in M_K(n,r)$. This functor has some agreeable properties, which we now describe.

(6.5c) *If $V \in M_K(N,r)$ then $eV = \sum V^\beta$, summed over those $\beta \in \Lambda(N,r)$ which lie in $\Lambda(n,r)^*$. Consequently the character Φ_{eV} (which is a symmetric polynomial over $\mathbb{Z}$ in n variables $X_1,\ldots,X_n$) is related to Φ_V (a polynomial in N variables $X_1,\ldots,X_N$) by the formula*

$$\Phi_{eV}(X_1,\ldots,X_n) = \Phi_V(X_1,\ldots,X_n,0,\ldots,0).$$

Proof. Suppose $\beta \in \Lambda(N,r)$, then $eV^\beta = e\xi_\beta V = V^\beta$ or zero, according as $\beta \in \Lambda(n,r)^*$ or not; the first statement of (6.5c) follows. The second statement comes from this, together with the definition of a character (see 3.4).

Next we look to see how our functor behaves on the modules $D_{\lambda,K}$, $V_{\lambda,K}$ and $F_{\lambda,K}$. For this we need a lemma.

(6.5d) Lemma. *Let $\lambda \in \Lambda^+(N,r)\backslash\Lambda(n,r)^*$ and $i \in I(n,r)$. Then the λ-tableau T_i is not standard.*

Proof. Clearly, $\lambda = (\lambda_1,\lambda_2,\ldots,\lambda_N)$ satisfies $\lambda_1 \geq \lambda_2 \geq \cdots \geq \lambda_N$, and since $\lambda \notin \Lambda(n,r)^*$ we must have $\lambda_{n+1} \neq 0$. So the λ-tableau T_i has at least $n+1$ places in its first column. But the entries in this column cannot all be distinct, since $i \in I(n,r)$. Therefore T_i is not standard (see (4.5a)).

(6.5e) Theorem. *Let $\lambda \in \Lambda^+(N,r)$, and let X_λ denote any one of $D_{\lambda,K}$, $V_{\lambda,K}$ or $F_{\lambda,K}$. Then $eX_\lambda \neq 0$ if and only if $\lambda \in \Lambda(n,r)^*$. In other words $eX_\lambda = 0$ if and only if λ has more than n non-zero parts.*

Proof. Suppose first that $\lambda \notin \Lambda(n,r)^*$, and that $\beta \in \Lambda(n,r)^*$. If $X_\lambda = D_{\lambda,K}$ or $V_{\lambda,K}$, then the dimension of X_λ^β is equal to the number of $i \in \beta$ such that the λ-tableau T_i is standard ((4.5a), (5.3b)). Since $i \in \beta$ implies $i \in I(n,r)$, we deduce from (6.5d) that $X_\lambda^\beta = 0$, and then from (6.5c) that $eX_\lambda = 0$. Since $F_{\lambda,K}$ is a factor module of $V_{\lambda,K}$, we have $eF_{\lambda,K} = 0$ also.

Conversely, suppose that $\lambda \in \Lambda(n,r)^*$. For any of the modules X_λ in question, we know that $X_\lambda^\lambda \neq 0$, hence $eX_\lambda \neq 0$ by (6.5c). This completes the proof of (6.5e).

If $\lambda \in \Lambda^+(N,r)$, then a necessary and sufficient condition that $\lambda \in \Lambda(n,r)^*$ is that $\lambda = \mu^*$ for some $\mu \in \Lambda^+(n,r)$. However, we know from (5.4b) that $\{\, F_{\lambda,K} : \lambda \in \Lambda^+(N,r) \,\}$ is a full set of irreducible modules in $M_K(N,r)$. So (6.2g) and the theorem just proved give the first statement in (6.5f) below.

(6.5f) *$\{eF_{\mu^*,K} : \mu \in \Lambda^+(n,r)\}$ is a full set of irreducible modules in $M_K(n,r)$. In fact $eF_{\mu^*,K} \cong F_{\mu,K}$ (isomorphism in $M_K(n,r)$). Hence there is the character formula, valid for all $\mu \in \Lambda^+(n,r)$, and for every characteristic p (including $p = 0$):*

$$\Phi_{\mu,p}(X_1, \ldots, X_n) = \Phi_{\mu^*,p}(X_1, \ldots, X_n, 0, \ldots, 0).$$

Proof. The second and third statements follow from the fact that $eF_{\mu^*,K}$ and $F_{\mu,K}$ are both irreducible modules in $M_K(n,r)$ having character with leading term $X_1^{\mu_1} \cdots X_n^{\mu_n}$ (see remark (i), 3.2).

Remarks. (i) A direct proof that $eF_{\mu^*,K} \cong F_{\mu,K}$ can be made along the following lines. Let $E(n)$ denote a K-space with basis $e_1, \ldots, e_n$, which we regard as subspace of a K-space $E(N)$ with basis $e_1, \ldots, e_N$. The inclusion $E(n)^{\otimes r} \subseteq E(N)^{\otimes r}$ can be shown to take $V_{\mu,K}$ to $eV_{\mu^*,K}$, and to induce an isomorphism $F_{\mu,K} \cong eF_{\mu^*,K}$. Similarly we may show that $D_{\mu,K} \cong eD_{\mu^*,K}$.

(ii) If $N \geq n \geq r$, then the map $\mu \mapsto \mu^*$ gives a bijection from $\Lambda^+(n,r)$ onto $\Lambda^+(N,r)$. For any $\lambda \in \Lambda^+(N,r)$, being a partition of r, can have at most r non-zero parts. Hence $\lambda = \mu^*$ for some uniquely defined element $\mu \in \Lambda^+(n,r)$. Then it follows from (6.5f) that the functor $d_{N,n} : M_K(N,r) \to M_K(n,r)$ induces a bijection between the sets of isomorphism classes of irreducible modules in these two categories. In fact we have the stronger result:

(6.5g) *If $N \geq n \geq r$, the functor $d = d_{N,n}$ defines an equivalence of categories. In particular $M_K(N,r) \simeq M_K(r,r)$ for all $N \geq r$.*

To prove (6.5g), we must produce a functor $h : M_K(n,r) \to M_K(N,r)$ such that hd, dh are naturally equivalent to the appropriate identity functors (see for example [10, p. 7]). We leave it to the reader to verify that we may take for h the functor described in 6.2, which in the present case takes each $W \in M_K(n,r)$ to

$$h(W) = S_K(N,r)e \otimes_{S_K(n,r)} W.$$

We might mention that (6.5g) has another formulation: If $N \geq n \geq r$, then the K-algebras $S_K(N,r)$ and $S_K(n,r)$ are *Morita equivalent* (see [10, p. 34]).

6.6 Application IV. Some theorems on decomposition numbers

In this section we first extend our "mod $S \to$ mod eSe" theory of 6.2 to a "modular" context, and prove a general result of T. Martins [41] on decomposition numbers. Then we apply this to the modular reduction $M_{\mathbb{Q}}(n,r) \to M_K(n,r)$ which was described in 2.5.

We start with a piece of notation. Let $\{ V_\lambda : \lambda \in \Lambda \}$ be a full set of irreducible modules in mod S, where S is an algebra over any field. If $V \in \text{mod } S$ and $\lambda \in \Lambda$, denote by $n_\lambda(V)$ the *composition multiplicity* of V_λ in V. That means, $n_\lambda(V)$ is the number of factors isomorphic to V_λ, in any composition series of V

(6.6a) $\quad V = V_0 \supset V_1 \supset \cdots \supset V_l = 0.$

Now let e be an idempotent of S, and define Λ', as in (6.2g), to be the set of those $\lambda \in \Lambda$ such that $eV_\lambda \neq 0$.

(6.6b) Lemma. *If $\lambda \in \Lambda'$, then $n_\lambda(eV) = n_\lambda(V)$, for any $V \in \text{mod } S$. Here $n_\lambda(eV)$ is the composition multiplicity of eV_λ in the eSe-module eV.*

Proof. From (6.6a) we get a series of eSe-modules

$$eV = eV_0 \supseteq eV_1 \supseteq \cdots \supseteq eV_l = 0.$$

By (6.2a), $e(V_{j-1}/V_j) \cong eV_{j-1}/eV_j$ (as eSe-modules), for $j = 1, \ldots, l$. Removing those terms eV_{j-1} for which $eV_{j-1} = eV_j$, i.e. for which V_{j-1}/V_j is isomorphic to V_π for some $\pi \in \Lambda \backslash \Lambda'$, we are left with a composition series for eV. The lemma follows at once.

Now let C, K be fields, and R be a subring of C with the properties

(i) R is a principal ideal domain (in particular, R contains 1_C),
(ii) C is the field of fractions of R, and
(iii) there is a ring-homomorphism $\pi : R \to K$.

Suppose S_C is a C-algebra with finite basis $\{u_1, \ldots, u_n\}$, such that the set $S_R = Ru_1 \oplus \cdots \oplus Ru_b$ contains the identity element of S_C and is multiplicatively closed: thus S_R is an "R-order" in S_C. Then $S_K = S_R \otimes K$ is a K-algebra with K-basis $\{u_1 \otimes 1_K, \ldots, u_n \otimes 1_K\}$ (here and below, $\otimes$ means $\otimes_R$, and K is regarded as R-module *via* π, i.e. $r \cdot k = \pi(r)k$, for $r \in R$, $k \in K$). R. Brauer's modular representation theory of algebras ([5]; see also [1, p. 111]) connects the categories mod S_C and mod S_K by the process of "modular reduction" (cf. 2.5), as follows. In each $V_C \in \text{mod } S_C$ can be found an "R-form" or "admissible R-lattice" V_R, i.e.

(i) V_R is the R-span of some C-basis of V_C, and
(ii) $S_R V_R \subseteq V_R$

(see [5, §6, p. 256], or [16, 48.1(iv), p. 299]). Then $V_K = V_R \otimes K$ can be regarded as left S_K-module; V_K is called a *modular reduction* of V_C. If

(6.6c) $\quad \{ V_{\lambda,C} : \lambda \in \Lambda \}, \quad \{ U_{\delta,K} : \delta \in \Delta \}$

are full sets of irreducible modules in mod S_C, mod S_K respectively, we define for each $\lambda \in \Lambda$, $\delta \in \Delta$ the *decomposition number* $d_{\lambda\delta} = d_{\lambda\delta}(S)$ to be the composition multiplicity $n_\delta(V_{\lambda,K})$ of $U_{\delta,K}$ in a modular reduction of $V_{\lambda,C}$.

Let $e = e_R$ be an idempotent in the ring S_R. Since $e \in S_C$, we can apply the theory of 6.2. By (6.2g) we get a full set of irreducible modules in mod eS_Ce, namely $\{ eV_{\lambda,C} : \lambda \in \Lambda' \}$, where $\Lambda' = \{ \lambda \in \Lambda : eV_{\lambda,C} \neq 0 \}$. Similarly, $e_K = e_R \otimes 1_K$ is an idempotent in the K-algebra $S_K = S_R \otimes K$, and so we get a full set of irreducible modules in mod $e_K S_K e_K$, namely $\{ e_K U_{\delta,K} : \delta \in \Delta' \}$, where $\Delta' = \{ \lambda \in \Delta : e_K U_{\delta,K} \neq 0 \}$.

Now eS_Re is an R-order in the C-algebra eS_Ce—this is because, as R-module, eS_Re is a direct summand of S_R. For the same reason, we can identify $eS_Re \otimes K$ with $e_K S_K e_K$. Therefore we have a process of modular reduction from mod eS_Ce to mod $e_K S_K e_K$; let us denote the corresponding decomposition numbers by $d_{\lambda\delta}(eSe)$. Of course these are defined only for $\lambda \in \Lambda'$, $\delta \in \Delta'$. The connection between these decomposition numbers, and the $d_{\lambda\delta}(S)$ defined previously, is very simple and satisfactory.

(6.6d) Theorem (T. Martins [41]). *Let $\lambda \in \Lambda'$, $\delta \in \Delta'$. Then*

$$d_{\lambda\delta}(S) = d_{\lambda\delta}(eSe).$$

Proof. Let V_R be an R-form of $V_C = V_{\lambda,C}$. Then eV_R is a direct summand of $V_R = eV_R \oplus (1-e)V_R$. This implies that eV_R is an R-form of the eS_Ce-module eV_C, and also that the $e_K S_K e_K$-module $eV_R \otimes K$ can be identified with $e_K V_K$, where $V_K = V_R \otimes K$. By (6.6b) the composition multiplicity of $e_K U_{\delta,K}$ in $e_K V_K$ is the same as the composition multiplicity of $U_{\delta,K}$ in V_K. This proves (6.6d).

For our applications of (6.6d) we take $C = \mathbb{Q}$ (field of rational numbers), $R = \mathbb{Z}$, K any infinite field of characteristic $p > 0$, and define $\pi : \mathbb{Z} \to K$ by $\pi(n) = n \cdot 1_K$ for all $n \in \mathbb{Z}$. Fix n, r and let $S = S_{\mathbb{Q}}(n,r)$, $S_{\mathbb{Z}} = S_{\mathbb{Z}}(n,r)$. Identify S_K with $S_K(n,r)$ by the isomorphism given in 2.3. Identify the categories mod $S_{\mathbb{Q}}$ and mod S_K with $M_{\mathbb{Q}}(n,r)$ and $M_K(n,r)$ respectively, as in 2.4.

Corresponding to the sets (6.6c) in the general case we take

$$\{ V_{\lambda,\mathbb{Q}} : \lambda \in \Lambda^+(n,r) \}, \quad \{ F_{\lambda,K} : \lambda \in \Lambda^+(n,r) \}.$$

(It happens in this case, these sets are indexed by the *same* set $\Lambda^+(n,r)$.) Denote the decomposition numbers by $d_{\lambda\mu} = d_{\lambda\mu}(\mathsf{GL}_n)$. These are the same numbers which appear in the formulae

$$\Phi_{\lambda,0}(X_1,\dots,X_n) = \sum_{\mu \in \Lambda^+(n,r)} d_{\lambda\mu} \Phi_{\lambda,p}(X_1,\dots,X_n)$$

of 3.5, remark (1).

(6.6e) Theorem. *Suppose that $N \geq n$, and let $\alpha \mapsto \alpha^*$ be the injective map from $\Lambda(n,r)$ into $\Lambda(N,r)$ given in 6.5. Then*

(i) $d_{\alpha\beta}(\mathsf{GL}_n) = d_{\alpha^*\beta^*}(\mathsf{GL}_N)$, *for any* $\alpha, \beta \in \Lambda^+(n,r)$.
(ii) $d_{\mu\nu}(\mathsf{GL}_N) = 0$, *for any* $\lambda \in \Lambda^+(N,r)\backslash\Lambda(n,r)^*$ *and* $\mu \in \Lambda(n,r)^*$.

Proof. (i) is a direct application of (6.6d). Take $S_{\mathbb{Q}} = S_{\mathbb{Q}}(N,r)$, etc., and let $e = e_{\mathbb{Z}}$ be the element of $S_{\mathbb{Q}}$ defined as in (6.5b). Clearly $e \in S_{\mathbb{Z}}$, and $e_K = e \otimes 1_K$ is the element of S_K defined by (6.5b). By (6.5e), the sets Λ', Δ' which appear in (6.6d) are both equal to $\Lambda^+(N,r) \cap \Lambda(n,r)^*$. So we take $\lambda = \alpha^*$, $\mu = \beta^*$ in (6.6d), and then use (6.5f).

(ii) Suppose $\lambda \notin \Lambda^+(n,r)^*$, then by (6.5e), $e_K V_{\lambda,K} = 0$. By (6.6b) the composition multiplicity of $F_{\mu,K}$ in $V_{\lambda,K}$ is the same as that of $eF_{\mu,K}$ in $e_K V_{\lambda,K}$, because $\mu \in \Lambda(n,r)^*$. In other words $d_{\lambda\mu}(\mathsf{GL}_N) = 0$, and the proof of (6.6e) is complete.

Remark. Part (i) of this theorem shows that, with fixed r and K, the decomposition numbers $d_{\lambda\mu}(\mathsf{GL}_n)$ *for all* n are contained in the matrix

(6.6f) $\quad (d_{\lambda\mu}(\mathsf{GL}_r))_{\lambda,\mu\in\Lambda(r)}$.

Here $\Lambda(r) = \Lambda^+(r,r)$, which can be identified with the set of all partitions $\lambda = (\lambda_1, \ldots, \lambda_r)$ of r. Assume first $n = r$ in (6.6e). Then $N \geq r$, and the map $\alpha \mapsto \alpha^*$ induces a bijection of $\Lambda(r)$ onto $\Lambda^+(N,r)$. So (6.6e)(i) shows that the decomposition matrix for GL_N is identical (up to this bijection) with (6.6f). Next take $N = r$. Then $n \leq r$ and the map $\alpha \mapsto \alpha^*$ takes $\Lambda^+(n,r)$ bijectively onto the set of those $\lambda \in \Lambda(r)$ which have not more than n non-zero parts. So the decomposition matrix for GL_n is identical (up to this bijection) with the submatrix of (6.6f) obtained by repressing all rows and columns which refer to partitions having more than n parts.

Theorem (6.6d) gives a simple proof of a theorem of James, which shows that the matrix of decomposition numbers for the symmetric group $G(r)$ is also a submatrix of (6.6f) — in this case we merely suppress those columns of (6.6f) which refer to partitions which are column p-singular. To see this, recall that we have full sets of irreducible $\mathbb{Q}G(r)$-, $KG(r)$-modules

$$\{\,\overline{S}_{\lambda,\mathbb{Q}} : \lambda \in \Lambda(r)\,\}, \quad \{\,D_\delta : \delta \in \Lambda^{(p)}(r)\,\},$$

respectively. Here $\Lambda^{(p)}(r)$ denotes the set of all column p-regular partitions of r. Let $d_{\lambda\delta}(G(r))$ denote the composition multiplicity of D_δ in $\overline{S}_{\lambda,K}$. Now we may apply (6.6d) with $S = S_{\mathbb{Q}}(n,r)$ ($r \leq n$), etc., $e = \xi_\omega \in S_{\mathbb{Q}}$, $e_K = \xi_\omega \in S_K$. We have $S_{\lambda,\mathbb{Q}} \cong eV_{\lambda,\mathbb{Q}}$ and $D_\delta \cong e_K F_{\delta,K}$ (see (6.3e), (6.4l) and the remarks preceding each). The result is as follows.

(6.6g) Theorem (James [28, Theorem 3.4]). *If* $r \leq n$, *then*

$$d_{\lambda\delta}(\mathsf{GL}_n) = d_{\lambda\delta}(G(r)),$$

for all $\lambda \in \Lambda(r)$ *and* $\delta \in \Lambda^{(p)}(r)$.

Tables showing the matrix (6.6f) for $2 \leq r \leq 6$, and $\operatorname{char} K = 2, 3$ are given in James's article [28].

Appendix on

Schensted correspondence and Littelmann paths

by K. Erdmann, J.A. Green and M. Schocker

A

Introduction

A.1 Preamble

These lectures describe some combinatorial properties of the set $I(n,r)$ of all "words" $i_1 i_2 \dots i_r$ of length r, whose "letters" $i_1, i_2, \dots, i_r$ are drawn from the "alphabet" $\underline{n} = \{1, \dots, n\}$. Clearly $I(n,r)$ is a finite set, with n^r elements.

Let $\Lambda(n,r)$ be the set of all vectors $\beta = (\beta_1, \dots, \beta_n)$ whose coefficients are non-negative integers satisfying $\sum_{\nu \in \underline{n}} \beta_\nu = r$. The elements $\beta \in \Lambda(n,r)$ are sometimes called *weights* (see section 3.1). Let $\Lambda^+(n,r)$ be the subset of $\Lambda(n,r)$ consisting of all β which are *dominant*, i.e. which satisfy $\beta_1 \geq \cdots \geq \beta_n (\geq 0)$. A dominant weight in this sense is often referred to as a *partition of r with no more than n parts.*

Example. $\Lambda^+(2,4) = \{(4,0), (3,1), (2,2)\}$.

The set $I(n,r)$ plays a humble rôle in the representation theory of the general linear group $\mathsf{GL}(n,K)$ (see section 2.6), because it indexes the basis $\{ v_i = v_{i_1} \otimes \cdots \otimes v_{i_r} \ : \ i \in I(n,r)\}$ of the r-fold tensor power $V^{\otimes r}$ of a vector space V of dimension n, with respect to a given basis $\{v_1, \dots, v_n\}$ of V.

But the present work is not based on linear algebra. We shall see that $I(n,r)$ has a rich combinatorial structure in its own right, based on two operations which may be performed on any word $i \in I(n,r)$; namely

(A.1a) the Robinson–Schensted algorithm, and

(A.1b) the application of maps $\tilde{e}_c$, $\tilde{f}_c$ which are essentially Littelmann's "*root operators*" e_α, f_α (see [35] and (A.3g)(2)).

Peter Littelmann uses the root operators as foundation of a remarkable theory [35], sometimes called the "path model" of the classical representation theory of GL_n; this is more combinatorial, and simpler in some ways, than the classical theory. Our work is an attempt to understand this "proto-representation theory" of GL_n.

A striking feature of Littelmann's theory is that it applies to arbitrary complex, symmetrizable Kac-Moody algebras. Our work, which applies only to sl_n, is therefore restricted to the special case of algebras of type A_{n-1}. But there is some advantage in this restriction; Littelmann's "paths" become "words", and we may work in the familiar combinatorial context of this set of lecture notes.

(A.1a) and (A.1b) will be described briefly in §A.2, §A.3, and discussed in more detail later.

A.2 The Robinson-Schensted algorithm

This algorithm (henceforth referred to as the Schensted process) turns a word $i \in I(n,r)$ into a triple $(\lambda(i), P(i), Q(i))$, where

(A.2a) $\lambda(i) = (\lambda_1(i), \ldots, \lambda_n(i))$ is a dominant weight; i.e. $\lambda(i)$ is a partition of r into at most n parts,

(A.2b) $P(i)$ is a standard tableau of "shape" $\lambda(i)$ (see section 4.2 and (4.5a)). The entries in the tableau $P(i)$ are the letters $i_1, i_2, \ldots, i_r$ in the word i, permuted in such a way that $P(i)$ is *standard*, i.e. so that the entries in each row of $P(i)$ are weakly increasing ($\leq$) from left to right, and the entries in each column are strictly increasing ($<$) from top to bottom.

(A.2c) $Q(i)$ is a standard tableau of "shape" $\lambda(i)$, whose entries are the integers $1, \ldots, r$ permuted in such a way that $Q(i)$ is standard[1].

Schensted calls $P(i)$ and $Q(i)$ the *P-symbol* and the *Q-symbol* (respectively) of the word i (see [46, p. 181]).

Schensted's rules which define $\lambda(i)$, $P(i)$ and $Q(i)$ will be given in §B.2. But it may be useful to look at the special case $r = 2$, where $\lambda(i)$, $P(i)$ and $Q(i)$ are easy to describe.

Take $r = 2$, and n any integer ≥ 2. A typical word in $I(n,2)$ is $i_1 i_2$. The only dominant weights are $(2,0,0,\ldots,0)$ and $(1,1,0,\ldots,0)$. The only values for $Q(i)$ are the tableaux $\begin{array}{|c|c|}\hline 1 & 2 \\ \hline \end{array}$ and $\begin{array}{|c|}\hline 1 \\ \hline 2 \\ \hline \end{array}$.

Table A.1 below (which is made using the rules in §B.2; see (B.4b)) produces a partition $I(n,2) = I_{\boxed{1|2}} \dot{\cup} I_{\begin{array}{|c|}\hline 1 \\ \hline 2 \\ \hline \end{array}}$, where

(A.2d) $I_{\boxed{1|2}} = \{\, i \in I(n,2) \,:\, i_1 \leq i_2 \,\}$ and $I_{\begin{array}{|c|}\hline 1 \\ \hline 2 \\ \hline \end{array}} = \{\, i \in I(n,2) \,:\, i_1 > i_2 \,\}$.

---

[1]Standard tableaux were first defined by A. Young [59] in his representation theory of the symmetric group $\mathsf{Sym}\{1, \ldots, r\}$. For this reason, standard tableaux are often called *Young tableaux*, or *generalized Young tableaux* [34].

From this table we see that the set $I_{\begin{array}{|c|c|}\hline 1&2\\\hline\end{array}}$ is the set of all $i \in I(n,2)$ such that $Q(i) = \begin{array}{|c|c|}\hline 1&2\\\hline\end{array}$, and $I_{\begin{array}{|c|}\hline 1\\\hline 2\\\hline\end{array}}$ is the set of all $i \in I(n,2)$ such that $Q(i) = \begin{array}{|c|}\hline 1\\\hline 2\\\hline\end{array}$. These sets are therefore the equivalence classes for the equivalence $\approx$ which we shall define in §A.4. The general case will be discussed in §C.1.

i	$\lambda(i)$	$P(i)$	$Q(i)$
$i_1 \leq i_2$	$(2,0,0,\ldots,0)$	$\begin{array}{\|c\|c\|}\hline i_1&i_2\\\hline\end{array}$	$\begin{array}{\|c\|c\|}\hline 1&2\\\hline\end{array}$
$i_1 > i_2$	$(1,1,0,\ldots,0)$	$\begin{array}{\|c\|}\hline i_2\\\hline i_1\\\hline\end{array}$	$\begin{array}{\|c\|}\hline 1\\\hline 2\\\hline\end{array}$

Table A.1. The Schensted process in case $n \geq r = 2$.

A.3 The operators $\tilde{\mathbf{e}}_\mathbf{c}$, $\tilde{\mathbf{f}}_\mathbf{c}$

Let $a, b \in \underline{n}$, $a \neq b$, and let $\alpha_{a,b} = (0,\ldots,0,1,0,\ldots,0,-1,0,\ldots,0)$ denote the element of $\mathbb{Z}^n$ which has 1, -1 at the places a, b respectively, and zero at all other places. These $n(n-1)$ vectors are called the *roots* of a system of type A_{n-1}. Define $\Sigma = \{\alpha_{1,2}, \alpha_{2,3}, \ldots, \alpha_{n-1,n}\}$. This is a subset of the set of all roots; its elements are called the *simple roots.*[2]

Choose an element $c \in \{1,2,\ldots,n-1\}$. To define Littelmann's operators $\tilde{e}_c$ and $\tilde{f}_c$ we need some preliminary definitions.

- Define the map $\omega = \omega_{c,c+1} : \underline{n} \to \mathbb{Z}$ by the rule $\omega(\nu) = 1$, -1 or zero, according as $\nu = c$, $\nu = c+1$, or $\nu \notin \{c, c+1\}$.
- Define the map $h_c^i : \{0,1,\ldots,r\} \to \mathbb{Z}$ by the rule:

  (A.3a) $h_c^i(0) = 0$, and $h_c^i(t) = \omega(i_1) + \cdots + \omega(i_t)$ for all $t \in \{1,\ldots,r\}$.

  This means for any $t \in \{1,\ldots,r\}$,

  (A.3b) $h_c^i(t)$ is the number of c's in the initial segment $i_1 i_2 \ldots i_t$ of the word i, minus the number of $c+1$'s in this segment.[3]

- Next let $M = M_c^i$ denote the largest of the integers $h_c^i(0), h_c^i(1), \ldots, h_c^i(r)$. Notice that M_c^i is always ≥ 0 since $h_c^i(0) = 0$.

[2] To read this Appendix, it is *not* necessary to know the theory of roots and root systems!

[3] h_c^i is sometimes called the *height function.*

- There may be several values of $t \in \{0, 1, \ldots, r\}$ such that $h_c^i(t) = M_c^i$; let $q = q_c^i$ be the least of these values, and let $\overline{q} = \overline{q}_c^i$ be the greatest.

(A.3c) Lemma. (i) *If* $q \neq 0$, *then* $i_q = c$.
(ii) *If* $\overline{q} \neq r$, *then* $i_{\overline{q}+1} = c + 1$.

Proof. (i) Suppose $q \neq 0$. We know that $h_c^i(q) = M$. Let $\mu = h_c^i(q-1)$. By (A.3a) $M = h_c^i(q) = \mu + \omega(i_q)$. The possible values for $\omega(i_q)$ are 1, -1 and 0. But if $\omega(i_q) = -1$ then $M = \mu - 1$, hence $\mu > M$ against the definition of M. If $\omega(i_q) = 0$, then $\mu = M$, against the definition of q, which says that q is the *least* value of t for which $h_c^i(t) = M$. Hence $\omega(i_q) = 1$, which implies that $i_q = c$. The proof of (ii) is similar, and is left to the reader.

(A.3d) Definition (see [35, §1]). With the notation given above, define maps $\tilde{e}_c, \tilde{f}_c : I(n,r) \to I(n,r) \cup \{\infty\}$ as follows.

(A.3e) If $M^i = 0$, define $\tilde{f}_c(i) = \infty$ (or say "$\tilde{f}_c(i)$ is undefined"). If $M^i \neq 0$, define $\tilde{f}_c(i)$ to be the word $s \in I(n,r)$ given by $s_t = i_t$ if $t \neq q$, and $s_q = c+1$.

(A.3f) If $M^i = h_c^i(r)$, define $\tilde{e}_c(i) = \infty$ (or say "$\tilde{e}_c(i)$ is undefined"). If $M^i \neq h_c^i(r)$, define $\tilde{e}_c(i)$ to be the word $s \in I(n,r)$ given by $s_t = i_t$ if $t \neq \overline{q}+1$, and $s_{\overline{q}+1} = c$.

(A.3g) Remarks.

(1) We have labelled these operators with the index c, rather than with the corresponding simple root $\alpha = \alpha_{c,c+1}$.
(2) Let $B : I(n,r) \to I(n,r)$ be the operator which turns each word $i_1 i_2 \ldots i_r$ into its "reverse" $i_r i_{r-1} \ldots i_2 i_1$. Then the maps just defined are related to Littelmann's "root operators" f_α, e_α (see [35, §1]) as follows: $\tilde{f}_c = B f_\alpha B$, $\tilde{e}_c = B e_\alpha B$.
(3) Let $i \in I(n,r)$. Then each of $\tilde{f}_c$, $\tilde{e}_c$ takes i either to ∞, or to a word which is identical to i except at one place. At this "critical place", $\tilde{f}_c(i)$ changes the entry from c to $c+1$, and $\tilde{e}_c(i)$ changes the entry from $c+1$ to c (see (A.3c)).
(4) The *weight* $\mathsf{wt}(i)$ of a word $i \in I(n,r)$ is the vector $\beta \in \mathbb{Z}^n$ defined as follows: for each $\nu \in \underline{n}$, β_ν is the number of places $\varrho \in \underline{r}$ for which $i_\varrho = \nu$ (see section 3.1). Then (3) shows that $\mathsf{wt}(\tilde{f}_c(i)) = \mathsf{wt}(i) - \alpha_{c,c+1}$, if $\tilde{f}_c(i) \neq \infty$. Similarly $\mathsf{wt}(\tilde{e}_c(i)) = \mathsf{wt}(i) + \alpha_{c,c+1}$, if $\tilde{e}_c(i) \neq \infty$.
(5) The maps $\tilde{f}_c$, $\tilde{e}_c$ are "inverse" to each other in the sense: if $\tilde{f}_c(i) \neq \infty$, then $\tilde{e}_c \tilde{f}_c(i) = i$, while if $\tilde{e}_c(i) \neq \infty$, then $\tilde{f}_c \tilde{e}_c(i) = i$.
(6) *Concatenation.* If $i \in I(n,r)$ and $j \in I(n,s)$, define the *concatenation* of i and j to be the word $i \mid j = (i_1, \ldots, i_r, j_1, \ldots, j_s) \in I(n, r+s)$. Then for any $c \in \{1, \ldots, n-1\}$ we have

$$\tilde{f}_c(i \mid j) = \begin{cases} \tilde{f}_c(i) \mid j & \text{if } M_c^i \geq h_c^i(r) + M_c^j, \text{ and} \\ i \mid \tilde{f}_c(j) & \text{if } M_c^i < h_c^i(r) + M_c^j, \end{cases}$$

and

$$\tilde{e}_c(i \mid j) = \begin{cases} \tilde{e}_c(i) \mid j & \text{if } M_c^i > h_c^i(r) + M_c^j, \text{ and} \\ i \mid \tilde{e}_c(j) & \text{if } M_c^i \leq h_c^i(r) + M_c^j. \end{cases}$$

All of the statements in (A.3g) are due (and in much greater generality) to Littelmann; see [35, §2]. However these statements are also easily verified directly from the definitions above.

As an example, we prove the second of the two statements in (A.3g)(5), namely

(5*) If $i \in I(n,r)$ and $c \in \{1,\ldots,n-1\}$ such that $\tilde{e}_c(i) \neq \infty$, then $\tilde{f}_c\tilde{e}_c(i) = i$.

Proof. To calculate $\tilde{e}_c(i)$, we first calculate the height function h_c^i. This function was defined in (A.3a): $h_c^i(0) = 0$, and $h_c^i(t) = \omega(i_1) + \cdots + \omega(i_t)$ for all $t \in \{1,\ldots,r\}$; it is given as the third line of table A.2 below. Let $M = M_c^i$; recall the definition of $\overline{q} = \overline{q}_c^i$ (see (A.3b) and (A.3c)), and notice that in our case $\overline{q} < r$, because $\tilde{e}_c(i) \neq \infty$. By (A.3c)(ii), $i_{\overline{q}+1} = c+1$. In the fourth row of table A.2 are inequalities (e.g. $h_c^i(t) \leq M$) which, taken together, express that $\overline{q}$ is the largest value of t such that $h_c^i(t) = M$.

t	0	1	2	$\cdots$	$\overline{q} = \overline{q}_c^i$	$\overline{q}+1$	$\cdots$	r
i_t		i_1	i_2	$\cdots$	$i_{\overline{q}}$	$i_{\overline{q}+1} = c+1$	$\cdots$	i_r
$h_c^i(t)$	0	$\omega(i_1)$	$\omega(i_1)+\omega(i_2)$	$\cdots$	$M = M_c^i$	$M-1$	$\cdots$	$h_c^i(r)$
			$\leq M$		M		$< M$	
s_t		i_1	i_2	$\cdots$	$i_{\overline{q}}$	c	$\cdots$	i_r
$h_c^s(t)$	0		$< M+1$			$M+1$	$\leq M+1$	
$\tilde{f}_c(s)_t$		i_1	i_2	$\cdots$	$i_{\overline{q}}$	$c+1$	$\cdots$	i_r

Table A.2. The height functions of i and $s = \tilde{e}_c(i)$.

According to definition (A.3f), the word $s = \tilde{e}_c(i)$ coincides with i except at the place $\overline{q}+1$. At this place $i_{\overline{q}+1} = c+1$, and $s_{\overline{q}+1} = c$. To calculate $\tilde{f}_c(s)$, we must know the function h_c^s. Clearly $h_c^s(t) = h_c^i(t)$ for all $t \in \{0,\ldots,\overline{q}\}$, and it is easy to see that $h_c^s(t) = h_c^i(t) + 2$ for all $t \in \{\overline{q}+1,\ldots,r\}$.

Now $h_c^i(\overline{q}+1) = h_c^i(\overline{q}) + \omega(i_{\overline{q}+1}) = M-1$, hence $h_c^s(\overline{q}+1) = M+1$. We may now check the inequalities in the penultimate line of table A.2. These show that $M_c^s = M+1 > 0$, and that $q_c^s = \overline{q}+1$. But then $\tilde{f}_c(s) = \tilde{f}_c(\tilde{e}_c(i)) = i$, as required.

A.4 What is to be done

The Schensted process associates to each $i \in I(n,r)$ a triple $(\lambda(i), P(i), Q(i))$. In §C.1 we shall define two equivalences on the set $I(n,r)$.

Definitions.

(A.4a) $i \sim j$ means that $P(i) = P(j)$, and

(A.4b) $i \approx j$ means that $Q(i) = Q(j)$.

Donald Knuth [34] introduced the relation $\sim$, and proved that $\sim$ is the equivalence on $I(n,r)$ generated by a collection of *basic moves* $i \to j$; see §C.3. The main result of these lectures is the following analogue to Knuth's theorem:

(A.4c) Theorem A. *Let $i, j \in I(n,r)$. Then $i \approx j$ if and only if there is a finite sequence of words (elements of $I(n,r)$):*

$$i(1), i(2), \dots, i(s)$$

such that $i(1) = i$, $i(s) = j$ and for each adjacent pair $i(\nu)$, $i(\nu+1)$ ***either*** *there exists an element $c \in \{1, \dots, n-1\}$ such that $\tilde{f}_c(i(\nu)) = i(\nu+1)$,* ***or*** *there exists an element $c \in \{1, \dots, n-1\}$ such that $\tilde{e}_c(i(\nu)) = i(\nu+1)$.*

Expressed less formally, Theorem A says that $\approx$ is the equivalence relation on $I(n,r)$ generated by a collection of basic moves $i \Rightarrow j$, where $i \Rightarrow j$ means that $\tilde{e}_c(i) = j$ for some $c \in \{1, 2, \dots, n-1\}$, or that $\tilde{f}_c(i) = j$.

Theorem A will be proved in §D.2. Chapter D also contains some notes on the representation theory of the "Littelmann algebra" $L(n,r)$, which is an analogue of the Schur algebra $S(n,r)$.

The proof of Theorem A depends on the following

(A.4d) Proposition B. *Let $c \in \{1, 2, \dots, n-1\}$. Then the operation $\tilde{f}_c$ commutes with the Schensted process, in the following sense:*

(A.4e) $KP(\tilde{f}_c(i)) = \tilde{f}_c(KP(i))$ for all $i \in I(n,r)$.

To explain the symbols KP which appear in (A.4e), we need the

Definition. Suppose given a standard tableau

$$P = \begin{matrix} y_{1,1} & y_{1,2} & \cdots & \cdots & y_{1,\lambda_1} \\ y_{2,1} & y_{2,2} & \cdots & y_{2,\lambda_2} & \\ \vdots & & & & \\ y_{m,1} & \cdots & y_{m,\lambda_m} & & \end{matrix}$$

of shape $\lambda \in \Lambda^+(n,r)$, with all its entries $y_{a,b} \in \underline{n}$. Define the "Knuth unwinding" of P to be the following word:

$$KP = y_{m,1} \cdots y_{m,\lambda_m} y_{m-1,1} \cdots y_{m,\lambda_{m-1}} \cdots y_{1,1} \cdots y_{1,\lambda_1} \, ;$$

see §C.2, or [34, page 173]. This is an element of $I(n,r)$, therefore we can apply the operators $\tilde{f}_c$, $\tilde{e}_c$ to it, and (A.4e) makes sense[4].

Proposition B will be proved in §C.4.

[4]Some authors identify the tableau P with the word KP, but to be cautious, we shall not make this identification in this Appendix.

B

The Schensted Process

B.1 Notations for tableaux

Choose a dominant weight $\lambda \in \Lambda^+(n,r)$. The *shape* of λ is the following subset of $\mathbb{Z} \times \mathbb{Z}$ (see, for example, section 4.2):

(B.1a) $[\lambda] := \{\, (a,b) \,:\, 1 \le a \le n,\ 1 \le b \le \lambda_a \,\}$.

A λ-*tableau*, or a *tableau of shape* λ, is a map $U : [\lambda] \to \mathbb{Z}$. We may think of U as a "partial matrix", with $U((a,b))$ as the *entry* in U at the place (a,b). These entries are elements of $\mathbb{Z}$, and U has entries only at the *places* $(a,b) \in [\lambda]$. The entry $U((a,b))$ is often denoted $u_{a,b}$.

We usually assume that a tableau $U = (u_{a,b})_{(a,b)\in[\lambda]}$ is *standard*; i.e. that each row (a) is weakly increasing from left to right: $u_{a,1} \le u_{a,2} \le \cdots \le u_{a,\lambda_a}$, and that each column (b) is strictly increasing from top to bottom: $u_{1,b} < u_{2,b} < \cdots < u_{\beta_b,b}$ (β_b is the length of column (b)). Notice that the latter condition implies that if the entries $u_{a,b}$ of a tableau U all lie in $\underline{n}$, then the number of rows of U cannot exceed n.

B.2 The map $\mathsf{Sch} : \mathbf{I(n,r)} \to \mathbf{T(n,r)}$

Let $T(n,r)$ be the set of all triples (λ, P, Q) such that $\lambda \in \Lambda^+(n,r)$, P is a λ-tableau whose entries are drawn from the set $\underline{n} = \{1,2,\ldots,n\}$, and Q is a λ-tableau whose entries are $1,2,\ldots,r$ in some order. (The r entries of Q are distinct; the r entries of P may include repetitions.)

The subject of this chapter is Schensted's map [46, pp. 180–181]

(B.2a) $\mathsf{Sch} : I(n,r) \to T(n,r)$.

We shall define Sch in §B.3, and prove in §B.6 that it is bijective [46, p. 182].

The map Sch is defined by induction on r. For $r = 1$, let

(B.2b) $\mathsf{Sch}(i) = \Big((1,0,\ldots,0), \boxed{i_1}, \boxed{1}\Big)$

for any one-letter word $i = i_1$ in $I(n,1)$. Note that $\boxed{i_1}$, $\boxed{1}$ are tableaux of shape $(1,0,\ldots,0)$, which means that each can be regarded as a 1×1 matrix.

From now on we assume $r > 1$.

Insertion. Fundamental for Schensted's work [46] is a process (or algorithm) which "inserts" a given element x_1 of $\underline{n}$ into a given tableau U. The result of this process is a tableau $U \leftarrow x_1$ whose entries are the entries of U (although perhaps in a different order), together with one extra entry x_1.

Example. Using the methods to be explained in §B.4, we shall show that if

$U = \begin{array}{|c|c|}\hline 1 & 2 \\ \hline 4 \\ \cline{1-1} \end{array}$ and $x_1 = 1$, then $U \leftarrow x_1$ is the tableau $\begin{array}{|c|c|}\hline 1 & 1 \\ \hline 2 \\ \cline{1-1} 4 \\ \cline{1-1} \end{array}$ (see (B.4b)).

The insertion process will be described in the next section; see (B.3b) and (B.3d).

Now suppose U is the middle term of an element (μ, U, V) of $T(n, r-1)$. As soon as we have calculated $P = U \leftarrow x_1$, we shall be able (see (B.3e)) to construct a dominant weight $\lambda \in \Lambda^+(n,r)$, and also a λ-tableau Q, such that (λ, P, Q) is an element of $T(n,r)$. We shall denote this element $(\mu, U, V) \leftarrow x_1$. See also [46, p. 181] and [34, pp. 712, 713].

The process provides the inductive step needed to define $\mathsf{Sch}(i)$ for any $i = i_1 i_2 \ldots i_{r-1} i_r \in I(n,r)$ $(r > 1)$, namely

(B.2c) Definition. $\mathsf{Sch}(i) := \mathsf{Sch}(i') \leftarrow i_r$, *where* $i' = i_1 i_2 \ldots i_{r-1}$.

Therefore we have a formula (which can also be used as a definition of $\mathsf{Sch}(i)$, see [46, p. 181]).

(B.2d) Formula. $\mathsf{Sch}(i) := (\cdots((\mathsf{Sch}(i_1) \leftarrow i_2) \leftarrow i_3) \quad \cdots \quad) \leftarrow i_r.$

B.3 Inserting a letter into a tableau

Suppose we have (1) an element (μ, U, V) of $T(n, r-1)$, where $r > 1$, and (2) an element x_1 of $\underline{n}$. In this section we define the element $(\mu, U, V) \leftarrow x_1$ of $T(n,r)$. To do this, we first define the tableau $U \leftarrow x_1$.

Schensted does this by modifying U row by row. For this purpose, it is convenient[1] to supplement each row (a) of U with two "virtual entries" $u_{a,0} = 0$ and $u_{a,\mu_a+1} = \infty$. Note that $(a,0)$ and $(a, \mu_a + 1)$ are *not* elements of $[\mu]$, and therefore $u_{a,0}$ and u_{a,μ_a+1} are not true entries in row (a).

Take any $y \in \underline{n}$. Even though y may not be equal to any of the entries $u_{a,k}$ of row (a), we may "position" y into row (a), using the following elementary lemma.

---

[1] See [34, p. 711].

(B.3a) Lemma. *For any $y \in \underline{n}$ and any $a \in \underline{n}$, there is a unique element $k(a) = k(a, y)$ in $\{1, 2, \ldots, \mu_a, \mu_a + 1\}$ such that $u_{a,k(a,y)-1} \le y < u_{a,k(a,y)}$.*

Proof. Define $k(a, y)$ to be the smallest $k \in \{1, 2, \ldots, \mu_a, \mu_a + 1\}$ such that $y < u_{a,k}$.

Example. Suppose $\mu_a = 5$, and that row (a) (including the virtual entries) is

$$(0)\ 2\ 2\ 2\ 3\ 7\ (\infty).$$

If $y = 2$, then $k(a) = 4$, because $u_{a,3} \le 2 < u_{a,4}$.
If $y = 1$, then $k(a) = 1$, because $u_{a,0} \le 1 < u_{a,1}$.
If $y = 4$, then $k(a) = 5$, because $u_{a,4} \le 4 < u_{a,5}$.
The situation $k(a, y) = \mu_a + 1 = 6$ occurs if and only if $u_{a,5} \le y < \infty$, that is, if and only if $y \ge 7$.

The insertion sequence. Let $\mu \in \Lambda^+(n, r)$, let U be a μ-tableau whose entries all lie in $\underline{n}$, and let $x_1 \in \underline{n}$. In order to define $P := U \leftarrow x_1$ first make the "insertion sequence"

(B.3b) x_1, $k(1)$, x_2, $k(2)$, ..., x_z, $k(z)$,

which contains all the data needed to construct $U \leftarrow x_1$.

Definition of the insertion sequence.

Step 1.
- x_1 is the given element of $\underline{n}$.
- $k(1)$ is the smallest $k \in \{1, \ldots, \mu_1, \mu_1 + 1\}$ such that $x_1 < u_{1,k}$. Equivalently, $k(1)$ is the unique element of $\{1, \ldots, \mu_1, \mu_1 + 1\}$ such that $u_{1,k(1)-1} \le x_1 < u_{1,k(1)}$. The case $k(1) = \mu_1 + 1$ occurs if and only if $u_{1,\mu(1)} \le x_1$ $(< x_{1,\mu(1)+1} = \infty)$, i.e. if and only if $x_1 \ge u_{1,\mu_1}$ (hence x_1 is $\ge$ all entries in row (1) of U).
- **If $k(1) = \mu_1 + 1$, the sequence is ended.**

Step 2. Now assume that $k(1) \ne \mu_1 + 1$. Then continue the definition of the insertion sequence.
- $x_2 := u_{1,k(1)}$.
- $k(2)$ is the smallest $k \in \{1, \ldots, \mu_2, \mu_2 + 1\}$ such that $x_2 < u_{2,k}$. Equivalently, $k(2)$ is the unique element of $\{1, \ldots, \mu_2, \mu_2 + 1\}$ such that $u_{2,k(2)-1} \le x_2 < u_{2,k(2)}$. The case $k(2) = \mu_2 + 1$ occurs if and only if $x_2 \ge u_{2,\mu(2)}$.
- **If $k(2) = \mu_2 + 1$, the sequence is ended.**

Step 3. Now assume that $k(2) \ne \mu_2 + 1$. Then continue
- $x_3 := u_{2,k(2)}$, etc.

Inductive Step. The general step is as follows: after x_{a-1} $(:= u_{a-2,k(a-2)})$ and $k(a-1)$ have been defined, then
- **If $k(a-1) = \mu_{a-1} + 1$, the sequence is ended.**

Now assume that $k(a-1) \ne \mu_{a-1} + 1$, and proceed to define

- $x_a := u_{a-1,k(a-1)}$;
- $k(a)$ is the least $k \in \{1, \ldots, \mu_a, \mu_a + 1\}$ such that $x_a < u_{a,k}$. Equivalently, $k(a)$ is the unique element of $\{1, \ldots, \mu_a, \mu_a + 1\}$ such that

(B.3c) $u_{a,k(a)-1} \leq x_a < u_{a,k(a)}$.

Definition of z. For each a such that $k(a-1) \neq \mu_{a-1} + 1$, (B.3c) shows that $x_a < u_{a,k(a)} = x_{a+1}$. Therefore the sequence $x_1 < x_2 < \cdots$ is finite ($x_1, x_2, \ldots$ are all elements of $\underline{n}$). Define z to be the largest element of $\underline{n}$ such that $k(z-1) \neq \mu_{z-1} + 1$. Then we must have $k(z) = \mu_z + 1$ (otherwise we could go on to define $k(z+1)$), and $u_{z,\mu_z} \leq x_z$ ($< \infty$), i.e. x_z is $\geq$ every entry in row (z) of U.

(B.3d) Definition of $\mathbf{U} \leftarrow \mathbf{x_1}$. Let $\lambda = \mu + \varepsilon_z$, where ε_z is the n-vector with 1 in place z, and zero at all other places. We shall show in (B.5b) that $\lambda \in \Lambda^+(n,r)$. Define $U \leftarrow x_1$ to be the λ-tableau $P = (p_{a,b})_{(a,b) \in [\lambda]}$ whose entry $p_{a,b}$ is identical with the corresponding entry $u_{a,b}$ of U, **except**

1° at the places $(a, k(a))$ for $a = 1, 2, \ldots, z-1$. At these places we define $p_{a,k(a)} = x_a$ (whereas $u_{a,k(a)} = x_{a+1}$), **and**

2° at place $(z, \mu_z + 1)$, where U has no entry, we define P to have entry $p_{z,\mu_z+1} = x_z$.

The shape of P is $\lambda = \mu + \varepsilon_z = (\mu_1, \ldots, \mu_{z-1}, \mu_z + 1, \mu_{z+1}, \ldots, \mu_n)$, because row (a) of P has the same length as row (a) of U, for all $a \neq z$, while the length of row (z) of P is one more than the length of row (z) of U.

Note that, for all $a > z$, the row (a) of P is identical to row (a) of U.

(B.3e) Definition of $(\mu, \mathbf{U}, \mathbf{V}) \leftarrow \mathbf{x_1}$. Let $(\mu, U, V) \in T(n, r-1)$, and let x_1 be an element of $\underline{n}$. Let P be the tableau $U \leftarrow x_1$ defined in (B.3d). Let λ be the weight $\mu + \varepsilon_z$. Define Q by enlarging the μ-tableau V, giving it a new entry r in place $(z, \mu_z + 1)$. Then $(\mu, U, V) \leftarrow x_1$ is by definition the triple (λ, P, Q).

(B.3f) Exercise. Prove that $k(1) \geq k(2) \geq \cdots \geq k(z)$ in any case.

[Hint. Let $a \in \{2, \ldots, z\}$. We must prove that $k(a) \leq k(a-1)$. By definition $k(a)$ lies in $\{1, \ldots, \mu_a + 1\}$, therefore $k(a) \leq \mu_a + 1$, which is $\leq k(a-1)$ if $k(a-1) \geq \mu_a + 1$. But if $k(a-1) < \mu_a + 1$, i.e. $k(a-1) \leq \mu_a$, then there exists an entry $u_{a,k(a-1)}$ in row (a) of U. Column standardness of U shows that $u_{a,k(a-1)} > u_{a-1,k(a-1)}$. Therefore $x_a = u_{a-1,k(a-1)} < u_{a,k(a-1)}$. But $k(a)$ is the least $k \in \{1, \ldots, \mu_a + 1\}$ such that $x_a < u_{a,k}$. It follows that $k(a) \leq k(a-1)$.]

Note. We have not yet proved that the triple (λ, P, Q) belongs to $T(n,r)$. For this we must show that $\lambda \in \Lambda^+(n,r)$, and that P, Q are standard. These things will be proved in §B.5, but we first look at some examples.

B.4 Examples of the Schensted process

The basic operation for the Schensted process is the insertion of a letter into a tableau. So suppose $r > 1$, let μ be an element of $\Lambda^+(n, r-1)$, let U be a μ-tableau, and let x_1 be any element of $\underline{n}$. We want to find the tableau $P = U \leftarrow x_1$.

The tableau $P = U \leftarrow x_1$ (see (B.3d)) can be made by modifying the rows $(1), (2), \dots$ of U, in turn. First "position" x_1 (which may or may not be equal to one of the entries of U) into row (1) of U (see Lemma (B.3a)). This means, find the (unique) element $k(1)$ such that $u_{1,k(1)-1} \leq x_1 < u_{1,k(1)}$. Assume that $k(1) \neq \mu_1 + 1$. Let $x_2 := u_{1,k(1)}$. Now let x_1 "bump"[2] x_2 into row (2), which means:

(i) change the entry $x_2 = u_{1,k(1)}$ in place $(1, k(1))$ to $x_1 = p_{1,k(1)}$, and then
(ii) "position" x_2 into row (2), that is: find the unique index $k(2)$ such that $u_{2,k(2)-1} \leq x_2 < u_{2,k(2)}$.

Then row (1) of U, changed by (i), is row (1) of P.

Now we are ready to change row (2) of U into row (2) of P. In general, when row $(a-1)$ of U has been changed into row $(a-1)$ of P, we define $x_a := k_{a-1,k(a-1)}$ and "bump" x_a into row (a). This process goes on until we reach row (z), where $k(z) = \mu_z + 1$. Then row (z) of P is made by adjoining an entry x_z to row (z) of U, in the new place $(z, \mu_z + 1)$ (which was not a place for U). All subsequent rows of P are the same as the corresponding rows of U.

It is sometimes better to use a slightly different "technology", to construct $U \leftarrow x_1$ from U. Here one makes the parameters $x_1, k(1), x_2, k(2), \dots$ as before, and records these on the tableau U; for each a we put bracketed (x_a) between the entries $u_{a,k(a-1)}$ and $u_{a,k(a)}$ of row (a). We do not change any of the entries of U. The resulting diagram (it is not a tableau in our sense) is called "U prepared for insertion of x_1". We pass from this diagram to $P = U \leftarrow x_1$ by replacing $\dots (x_a)\ x_{a+1} \dots$ by $\dots x_a \dots$, for each $a \in \{1, \dots, z-1\}$; for row (z), replace $\dots u_{z,\mu_z}\ (x_z)$ by $\dots u_{z,\mu_z}\ x_z$.

(B.4a) Example. Suppose we want to insert $x_1 = 2$ into the tableau U shown in the left-hand column of table B.1 below. The second column shows U "prepared" for this insertion. This means that we have put bracketed (x_a) between the entries $u_{a,k(a)-1}, u_{a,k(a)}$ for each $a = 1, 2, \dots, z$; we have not yet changed any of the entries of U. Once U has been prepared, make $P = P(i)$ from U by changing the entry $u_{a,k(a)} = x_{a+1}$ to $p_{a,k(a)} = x_a$, for all $a = 1, 2, \dots, z-1$, i.e. the term x_a in the bracketed (x_a) replaces its right-hand neighbour $x_{a+1} = u_{a,k(a)}$. This procedure determines row (z), namely it is the first row where (x_z) does not have a right-hand neighbour. The row (z) of P, is

[2]The verb "bump" was introduced, in this context, by Knuth [34, p. 713]. Alternatives would be "dump", or even "jump".

found by replacing (x_z) by x_z. In the example shown, we have $x_1 = 2$, $x_2 = 3$, $z = 2$, $k(1) = 5$, $k(2) = 3$. Note that 3 in row (1) of U, is bumped into row (2).

U	U prepared for insertion of 2	$P = U \leftarrow 2$
1 1 2 2 3 3 3 4	1 1 2 2 (2) 3 3 3 (3) 4	1 1 2 2 2 3 3 3 4

Table B.1. Example for the insertion process.

(B.4b) Example. Consider two "extreme" possibilities.

(i) It can happen that, for some a, the element $x_a < u_{a,1}$. In that case $k(a) = 1$, and (B.3a) shows that $0 \leq x_a < u_{a,1}$. When U is "prepared" as above, row (a) looks like this:

$$(x_a)\ u_{a,1}\ u_{a,2} \cdots u_{a,\mu_a} .$$

(ii) It can happen that, for some a, we have $k(a) \neq \mu_a + 1$, and $\mu_{a+1} = 0$. This means that we must "bump" x_{a+1} $(= u_{a,k(a)})$ into an empty row $(a+1)$. We have $0 < x_{a+1} < \infty$, which allows us to say that $k(a+1) = 1$. Then $k(a+1) = \mu_{a+1} + 1$. So $a + 1 = z$, and P has an entry x_{a+1} in place $(a+1, 1)$.

The following example illustrates both possibilities (i) and (ii). Suppose we insert $x_1 = 1$ into the tableau $U = \begin{array}{|c|c|}\hline 1 & 2 \\ \hline 4 \\ \cline{1-1}\end{array}$. Prepared for this insertion, U becomes $\begin{array}{|c|c|c|}\hline 1 & (1) & 2 \\ \hline (2) & 4 \\ \cline{1-2} (4) \\ \cline{1-1}\end{array}$. Hence $P = U \leftarrow 1 = \begin{array}{|c|c|}\hline 1 & 1 \\ \hline 2 \\ \cline{1-1} 4 \\ \cline{1-1}\end{array}$. In this example x_1, x_2, x_3 are 1, 2, 4, respectively, $z = 3$; and $k(1) = 2$, $k(2) = 1$, $k(3) = 1$.

We are now in a position to calculate the P-symbol $P(i)$ and the Q-symbol $Q(i)$ of a given word $i \in I(n,r)$. To find $P(i) = P(i_1 i_2, \cdots i_r)$, we must calculate, successively, the P-symbols of the words i_1, $i_1 i_2$, $\ldots$, $i_1 i_2 \cdots i_r$, starting with $P(i_1) = \boxed{i_1}$, and using the insertion process

$$P(i_1 i_2 \cdots i_t) = P(i_1 i_2 \cdots i_{t-1}) \leftarrow i_t .$$

It follows that the entries of $P(i)$ are the entries $i_1, \ldots, i_r$ of i, in some order. The construction of $Q(i)$ is different, $Q(i_1 \cdots i_t)$ is *not* made by inserting i_t into $Q(i_1 \cdots i_{t-1})$; the construction follows definition (B.3e), see Example (B.4c) below.

(B.4c) Example. Calculate $P(i)$, where $i = 1\,4\,2\,1\,2$. Calculate also $\lambda(i)$ and $Q(i)$.

First we must work out the successive tableaux $P_t(i) = P(i_1 \ldots i_t)$, for $t = 1, 2, \ldots, 5$. We find (the operator $\xrightarrow{y}$ means "insert y into the tableau on the left")

$$P_1(i) = \begin{array}{|c|}\hline 1 \\ \hline\end{array} \xrightarrow{4} P_2(i) = \begin{array}{|c|c|}\hline 1 & 4 \\ \hline\end{array} \xrightarrow{2} P_3(i) = \begin{array}{|c|c|}\hline 1 & 2 \\ \hline 4 \\ \cline{1-1}\end{array}$$

$$\xrightarrow{1} P_4(i) = \begin{array}{|c|c|}\hline 1 & 1 \\ \hline 2 \\ \cline{1-1} 4 \\ \cline{1-1}\end{array} \xrightarrow{2} P_5(i) = P(i) = \begin{array}{|c|c|c|}\hline 1 & 1 & 2 \\ \hline 2 \\ \cline{1-1} 4 \\ \cline{1-1}\end{array}.$$

At the same time we get the dominant weight at each stage, namely $\lambda(i_1 \cdots i_t)$ is just the shape of the tableau $P_t(i)$. In particular, $\lambda(i) = (3, 1, 1, 0, \ldots, 0)$. Now we make the tableaux $Q_t(i) = Q(i_1 \ldots i_t)$ as follows: if we know $Q_{t-1}(i)$, then $Q_t(i)$ is got by putting "t" in the place which was new, when $P_t(i)$ was constructed from $P_{t-1}(i)$. Thus

$$Q_1(i) = \begin{array}{|c|}\hline 1 \\ \hline\end{array}, \quad Q_2(i) = \begin{array}{|c|c|}\hline 1 & 2 \\ \hline\end{array}, \quad Q_3(i) = \begin{array}{|c|c|}\hline 1 & 2 \\ \hline 3 \\ \cline{1-1}\end{array},$$

$$Q_4(i) = \begin{array}{|c|c|}\hline 1 & 2 \\ \hline 3 \\ \cline{1-1} 4 \\ \cline{1-1}\end{array}, \quad Q_5(i) = Q(i) = \begin{array}{|c|c|c|}\hline 1 & 2 & 5 \\ \hline 3 \\ \cline{1-1} 4 \\ \cline{1-1}\end{array}.$$

(B.4d) Example. Calculate $\lambda(i)$, $P(i)$, $Q(i)$ for any word $i = i_1 i_2 \in I(n, 2)$, and so verify the table given in §A.2.

To find $P(i)$, we must insert i_2 into the tableau $U = \begin{array}{|c|}\hline i_1 \\ \hline\end{array}$. When U is prepared for this insertion, it becomes $\begin{array}{|c|c|}\hline i_1 & (i_2) \\ \hline\end{array}$ in case $i_1 \leq i_2$, and it becomes $\begin{array}{|c|c|}\hline (i_2) & i_1 \\ \hline (i_1) \\ \cline{1-1}\end{array}$ in case $i_1 > i_2$. Therefore $P(i) = \begin{array}{|c|c|}\hline i_1 & i_2 \\ \hline\end{array}$ in case $i_1 \leq i_2$, and $P(i) = \begin{array}{|c|}\hline i_2 \\ \hline i_1 \\ \hline\end{array}$ in case $i_1 > i_2$. It follows that $\lambda(i)$ is $(2, 0, 0, \ldots, 0)$ or $(1, 1, 0, \ldots, 0)$, in these respective cases.

To find $Q(i)$, we must add 2 to $V = Q(i_1) = \begin{array}{|c|}\hline 1 \\ \hline\end{array}$ in the place $(z, \mu_z + 1)$. In the case $i_1 \leq i_2$, we have $z = 1$, so $Q(i) = \begin{array}{|c|c|}\hline 1 & 2 \\ \hline\end{array}$. In case $i_1 > i_2$, we have $z = 2$, so $Q(i) = \begin{array}{|c|}\hline 1 \\ \hline 2 \\ \hline\end{array}$.

B.5 Proof that $(\mu, U, V) \leftarrow x_1$ belongs to $T(n, r)$

We keep the notations of §§B.2, B.3, B.4. Suppose that $r > 1$ and that $(\mu, U, V) \in T(n, r-1)$. Let $x_1 \in \underline{n}$. The triple $(\lambda, P, Q) = (\mu, U, V) \leftarrow x_1$ is defined in (B.3e). In this section we shall prove that $(\lambda, P, Q) \in T(n, r)$. For this we must show that $\lambda \in \Lambda^+(n, r)$, and that P, Q are both standard.

Write $u_{a,b}$ for the (a,b)-entry of U, and $p_{a,b}$ for the (a,b)-entry of P.

(B.5a) Proposition. *The weight*

$$\lambda = \mu + \varepsilon_z = (\mu_1, \ldots, \mu_{z-1}, \mu_z + 1, \mu_{z+1}, \ldots, \mu_n)$$

is dominant. It follows that Q is standard.

Proof. We already know that $\mu_1 \geq \cdots \geq \mu_{z-1} \geq \mu_z \geq \mu_{z+1} \geq \cdots \geq \mu_n$ because μ is dominant. If λ is not dominant, it must be that $\mu_{z-1} = \mu_z$. But this leads to a contradiction. We know that $x_z \geq u_{z,\mu_z}$, and that $u_{z,\mu_z} > u_{z-1,\mu_z}$ because U is standard. But $u_{z-1,\mu_z} \geq u_{z-1,k(z-1)} = x_z$ (the last equality is the definition of x_z), and putting these inequalities together gives the contradiction $x_z > x_z$.

By the definition (B.3e), Q is made by adding an entry r at the end of row (z) of V. It is clear that Q is a standard λ-tableau, whose entries are $1, 2, \ldots, r$ in some order.

(B.5b) Proposition. *The λ-tableau P is standard.*

Proof. First we shall show that P is "row standard", i.e. that

(i) $p_{a,h-1} \leq p_{a,h}$

for all adjacent pairs $(a, h-1)$, (a, h) of places in any row (a) of $[\lambda]$.

If $(a, k(a))$ is not one of $(a, h-1)$, (a, h) then by (B.5a) $p_{a,h} = u_{a,h}$ and $p_{a,h-1} = u_{a,h-1}$, therefore (i) follows from the corresponding fact for row (a) of U. If $(a, h) = (a, k(a))$ then $p_{a,h-1} = u_{a,h-1} \leq x_a$, and $x_a = p_{a,k(a)} = p_{a,h}$; thus (i) holds. There remains the case $(a, h-1) = (a, k(a))$. Then (i) says $p_{a,k(a)} \leq p_{a,k(a)+1}$. But $p_{a,k(a)} = x_a$ and $p_{a,k(a)+1} = u_{a,k(a)+1}$. Thus (i) follows from $x_a < x_{a+1} = u_{a,k(a)} \leq u_{a,k(a)+1}$.

To complete the proof of Proposition (B.5b), we must show that P is "column standard", i.e. that if (a, h) and $(a+1, h)$ are adjacent places in the same column of $[\lambda]$, then

(ii) $p_{a+1,h} > p_{a,h}$.

If $h \neq k(a)$ and $h \neq k(a+1)$ then $p_{a,h} = u_{a,h}$ and $p_{a+1,h} = u_{a+1,h}$, hence (ii) follows from $u_{a+1,h} > u_{a,h}$, which holds because U is column standard.

If $h = k(a)$, $h \neq k(a+1)$ then $u_{a,h} = u_{a,k(a)} = x_{a+1} > x_a$. But $u_{a+1,k(a)} > u_{a,k(a)}$ because U is column standard. Therefore $p_{a+1,k(a)} = u_{a+1,k(a)} > x_a = p_{a,k(a)}$, which proves (ii) in this case.

Now suppose that $h = k(a+1)$, $h \neq k(a)$. Then $p_{a+1,k(a+1)} = x_{a+1}$, and $p_{a,k(a+1)} = u_{a,k(a+1)}$. In place $(a, k(a))$ of P we have x_a (see (B.3d)). Since P is "row standard" (just proved, above) and $k(a+1) \leq k(a)$ (see (B.3f)) we have $u_{a,k(a+1)} \leq x_a$; also $x_a < x_{a+1}$ by (B.3b). So $p_{a,k(a+1)} = u_{a,k(a+1)} \leq x_a < x_{a+1} = p_{a+1,k(a+1)}$. This proves (ii) in case $h = k(a+1)$, $h \neq k(a)$.

There remains only the case $h = k(a) = k(a+1)$. In this case $p_{a+1,h} = p_{a+1,k(a+1)} = x_{a+1}$ and $p_{a,h} = p_{a,k(a)} = x_a$. But $x_{a+1} > x_a$, therefore (ii) holds. The proof of Proposition (B.5b) is now complete.

B.6 The inverse Schensted process

This section and the next are devoted to Schensted's fundamental

(B.6a) Theorem (see [46, p. 182]; [34, pp. 715–716]). *The map*

$$\mathsf{Sch} : I(n,r) \to T(n,r)$$

is bijective.

This will be proved by constructing a map $\mathsf{M} : T(n,r) \to I(n,r)$ which is a two-sided inverse to Sch (see (B.7b)).

If $r = 1$, it is easy to make a map M inverse to Sch. The only element in $\Lambda^+(n,1)$ is $\lambda = (1,0,\dots,0)$, hence any element in $T(n,1)$ has the form $(\lambda, \boxed{x}, \boxed{1})$ for some $x \in \underline{n}$. We define $\mathsf{M}((\lambda, \boxed{x}, \boxed{1}))$ to be x (regarded as a 1-letter word). By (B.2b), $\mathsf{Sch}(x) = (\lambda, \boxed{x}, \boxed{1})$. It is easy to check now that $\mathsf{M} : T(n,1) \to I(n,1)$ is a two-sided inverse to $\mathsf{Sch} : I(n,1) \to T(n,1)$. It follows that Sch is bijective in case $r = 1$.

From now on in this section, assume that $r > 1$.

The process given in §B.3 delivers a map, which we call *insertion*,

(B.6b) $\mathsf{J} : T(n, r-1) \times \underline{n} \to T(n,r)$,

which takes a pair $((\mu, U, V), x_1)$ to the element $(\mu, U, V) \leftarrow x_1$ of $T(n,r)$.

Next define another map, called *extrusion*,

(B.6c) $\mathsf{E} : T(n,r) \to T(n, r-1) \times \underline{n}$.

To make E, we need an "inverse Schensted process", which will turn any $(\lambda, P, Q) \in T(n,r)$ into a pair consisting of a triple $(\mu, U, V) \in T(n, r-1)$ and an element $w_1 \in \underline{n}$.

How to define E. Let $(\lambda, P, Q) \in T(n,r)$. Let $(a,b) \in [\lambda]$ be the (unique) place where $q_{a,b} = r$. Since Q is standard, r must be at the end of its row. Therefore if $a = z$, then b must be λ_z, so that $q_{z,\lambda_z} = r$. But r is also at the end of its column, which implies that $\lambda_z > \lambda_{z+1}$. This proves

(B.6d) The weight $\mu = \lambda - \varepsilon_z$ is dominant. Hence $\mu \in \Lambda^+(n, r-1)$.

Definition of the extrusion sequence. We shall next define the *extrusion sequence*

(B.6e) $l(z)$, w_z, $l(z-1)$, w_{z-1}, ..., $l(1)$, w_1.

To make this sequence, we must know Q (which determines z) as well[3] as the λ-tableau P.

Step 1.
- $l(z) = \lambda_z$;
- $w_z := p_{z,\lambda_z}$ (this is the entry in P, at the place (z, λ_z) where Q has entry r).

Step 2.
- $l(z-1)$ is the largest $l \in \{1, 2, \ldots, \lambda_{z-1}\}$ such that $p_{z-1,l} < w_z$. Equivalently, $l(z-1)$ is the unique element in $\{1, 2, \ldots, \lambda_{z-1}\}$ such that $p_{z-1,l(z-1)} < w_z \leq p_{z-1,l(z-1)+1}$.
- $w_{z-1} := p_{z-1,l(z-1)}$.

Inductive Step. When $l(a+1)$ and $w_{a+1} := p_{a+1,l(a+1)}$ have been defined, we go on to define
- $l(a)$ is the largest $l \in \{1, \ldots, \lambda_a\}$ such that $p_{a,l} < w_{a+1}$. Equivalently, $l(a)$ is the unique element in $\{1, \ldots, \lambda_a\}$ such that

(B.6f) $p_{a,l(a)} < w_{a+1} \leq p_{a,l(a)+1}$.

- $w_a := p_{a,l(a)}$.

Note that if $a < z$ there is always at least one $l \in \{1, 2, \ldots, \lambda_a\}$ such that $p_{a,l} < w_{a+1}$, namely $l = l(a+1)$; this is because P is column standard, hence $p_{a,l(a+1)} < p_{a+1,l(a+1)} = w_{a+1}$. So the extrusion sequence (B.6e) always ends with ..., $l(1)$, w_1.

Final Step. The last two terms are as follows.
- $l(1)$ is the largest $l \in \{1, 2, \ldots, \lambda_1\}$ such that $p_{1,l} < w_2$, and
- $w_1 := p_{1,l(1)}$.

We say that the element $w_1 \in \underline{n}$ has been "extruded"[4] from P (or more precisely from the given element (λ, P, Q) in $T(n, r)$). But the extrusion process also defines an element (μ, U, V), see below.

(B.6g) Definition of E. Let $(\lambda, P, Q) \in T(n, r)$. Then $\mathsf{E}((\lambda, P, Q))$ is the pair $((\mu, U, V), w_1)$, where

- $\mu = \lambda - \varepsilon_z = (\lambda_1, \ldots, \lambda_{z-1}, \lambda_z - 1, \lambda_{z+1}, \ldots, \lambda_n)$,

[3] To define $P = U \leftarrow x_1$, we did not need to know Q, because z is defined by the insertion sequence; see (B.3c), (B.3d).

[4] In the way that a small amount (w_1) of toothpaste is *extruded* from its tube (λ, P, Q).

- V is Q, with the entry $q_{z,\lambda_z} = r$ removed,
- w_1 is the extruded element of $\underline{n}$ defined above, and
- $U = (u_{a,b})_{(a,b)\in[\mu]}$ is the following μ-tableau: $u_{a,b} = p_{a,b}$ for all $(a,b) \in [\mu]$, **except**
  1° at the places $(a, l(a))$ for $a = 1, 2, \ldots, z-1$. At these places we define $u_{a,l(a)} = w_{a+1}$ (whereas $p_{a,l(a)} = w_a$), **and**
  2° there is no entry in U at the place (z, λ_z), because U is a μ-tableau and $(z, \lambda_z) \notin [\mu]$.

To complete the definition of E, the following lemma is required.

(B.6h) Lemma. *The triple* (μ, U, V) *belongs to* $T(n, r-1)$.

Proof. From (B.6d) we know that $\mu \in \Lambda^+(n, r-1)$. It is clear that V is a μ-tableau whose entries are $1, 2, \ldots, r-1$, in some order. It remains only to show that U is standard. The proof of this is very similar to that of Proposition (B.5b), and we leave to the reader.

(B.6i) Proposition. *The maps* J, E *are inverse to each other.*

Proof. We shall first prove that

(i) $\mathsf{E} \circ \mathsf{J} = \mathrm{id}_{T(n,r-1)\times\underline{n}}$.

Take any element $((\mu, U, V), x_1)$ in $T(n, r-1) \times \underline{n}$. Let

(B.6j) $x_1,\ k(1),\ \ldots,\ x_z,\ k(z),$

be the insertion sequence used to define $(\lambda, P, Q) = \mathsf{J}((\mu, U, V), x_1) = (\mu, U, V) \leftarrow x_1$. Here z is such that $k(z) = \mu_z + 1 = \lambda_z$, where $\lambda = \mu + \varepsilon_z$. Note that Q has r in place (z, λ_z), and $p_{z,\lambda_z} = x_z$ (see (B.3d) and (B.3e)).

To prove (i) it is enough to show that $\mathsf{E}((\lambda, P, Q)) = ((\mu, U, V), x_1)$. Now $\mathsf{E}((\lambda, P, Q))$ is determined by the extrusion sequence (see (B.6e))

(B.6k) $l(z),\ w_z,\ l(z-1),\ w_{z-1},\ \ldots,\ l(1),\ w_1.$

The "z" which appears in (B.6k) indexes the row of Q which contains the entry r, see (B.6d). Therefore this "z" is the same as the z in (B.6j). From (B.6e) we have $l(z) = \lambda_z$ and $w_z = p_{z,\lambda_z}$. But from the definition of P, $p_{z,\lambda_z} = p_{z,\mu_z+1} = x_z$. Therefore

(B.6l) $l(z) = k(z)$ *and* $w_z = x_z$.

Our ambition is to prove

(B.6m) $l(a) = k(a)$ *and* $w_a = x_a$

for all $a \in \{z, z-1, \ldots, 1\}$. Suppose $a < z$ and that (using "upward" induction)

(B.6n) $l(a+1) = k(a+1)$ *and* $w_{a+1} = x_{a+1}$.

By (B.3c), there holds $x_a < x_{a+1} = u_{a,k(a)} \le u_{a,k(a)+1}$. However the definitions in §B.3 show that $x_a = p_{a,k(a)}$, and (B.6n) gives $w_{a+1} = x_{a+1}$. So $x_a = p_{a,k(a)} < w_{a+1} \le u_{a,k(a)+1} = p_{a,k(a)+1}$. But comparing this with (B.6f), we see that $l(a) = k(a)$. Hence $w_a = p_{a,l(a)} = p_{a,k(a)} = x_a$; thus (B.6m) holds for all a. In particular, $w_1 = x_1$, and we find easily that

$$\mathsf{E}(\mathsf{J}(((\mu, U, V), x_1)))) = \mathsf{E}((\lambda, P, Q)) = ((\mu, U, V), x_1);$$

in other words we have proved (i).

To complete the proof of (B.6i) we must prove

(ii) $\mathsf{J} \circ \mathsf{E} = \mathsf{id}_{T(n,r)}$.

Take any element $(\lambda, P, Q) \in T(n,r)$. Let (B.6k) be the extrusion sequence which defines $\mathsf{E}((\lambda, P, Q)) = ((\mu, U, V), w_1)$ (see (B.6g)). Let

(B.6o) $(w_1 =)\ x_1,\ k(1),\ x_2,\ k(2),\ \ldots,\ x_z,\ k(z)$

be the insertion sequence which defines $\mathsf{J}((\mu, U, V), w_1)$. In order to prove (ii) we must show that $\mathsf{J}((\mu, U, V), w_1) = (\lambda, P, Q)$.

The first step is to prove

(B.6p) $w_a = x_a$

for all $a \in \{1, 2, \ldots, z\}$. This holds for $a = 1$, by definition. Suppose that (B.6p) holds for some a. By (B.6f), $l(a)$ is the unique element of $\{1, 2, , \ldots, z\}$ such that

(B.6q) $p_{a,l(a)} = w_a < w_{a+1} \le p_{a,l(a)+1}$.

From this follows that $p_{a,l(a)-1} \le w_a < w_{a+1}$. Hence, using the definition (B.6g) of U, we have $u_{a,l(a)-1} \le w_a < u_{a,l(a)}$, and since $w_a = x_a$, there holds

$$u_{a,l(a)-1} \le x_a < u_{a,l(a)}\,.$$

However this proves that $l(a) = k(a)$, from (B.3c). Consequently $w_{a+1} = p_{a,l(a)} = p_{a,k(a)} = x_{a+1}$ (see (B.3b); we are here using the insertion of x_1 into (μ, U, V)). Now we can prove, by induction on a, that

(B.6r) $w_a = x_a$ and $l(a) = k(a)$, for all $a \in \{1, 2, \ldots, z\}$.

Using (B.6r) and the definitions (B.3d) and (B.3e) (applied to the insertion of x_1 into (μ, U, V)), it is quite easy to show that $\mathsf{J}((\mu, U, V), w_1) = (\lambda, P, Q)$. This concludes the proof of Proposition (B.6i).

B.7 The ladder

We shall define a map $\mathsf{M} : T(n,r) \to I(n,r)$ inverse to $\mathsf{Sch} : I(n,r) \to T(n,r)$, and hence prove Schensted's Theorem (B.6a). M will be given as the product of maps $\mathsf{E}_0, \mathsf{E}_1, \ldots, \mathsf{E}_{r-1}$ displayed in table B.2 ("The ladder") below.

Set	Typical element of set
$I(n,r)$	$i_1 i_2 \ldots i_s i_{s+1} \ldots i_{r-1} i_r$
$\mathsf{J}_1 \downarrow \quad \uparrow \mathsf{E}_{r-1}$	
$T(n,1) \times I(n,r-1)$	$\big((\lambda_1, P_1, Q_1), i_2 \ldots i_s i_{s+1} \ldots i_{r-1} i_r\big)$
$\mathsf{J}_2 \downarrow \quad \uparrow \mathsf{E}_{r-2}$	
$\vdots$	$\vdots$
$\mathsf{J}_{s-1} \downarrow \quad \uparrow \mathsf{E}_{r-s+1}$	
$T(n,s-1) \times I(n,r-s+1)$	$\big((\lambda_{s-1}, P_{s-1}, Q_{s-1}), i_s i_{s+1} \ldots i_{r-1} i_r\big)$
$\mathsf{J}_s \downarrow \quad \uparrow \mathsf{E}_{r-s}$	
$T(n,s) \times I(n,r-s)$	$\big((\lambda_s, P_s, Q_s), i_{s+1} \ldots i_{r-1} i_r\big)$
$\mathsf{J}_{s+1} \downarrow \quad \uparrow \mathsf{E}_{r-s-1}$	
$\vdots$	$\vdots$
$\mathsf{J}_{r-1} \downarrow \quad \uparrow \mathsf{E}_1$	
$T(n,r-1) \times I(n,1)$	$\big((\lambda_{r-1}, P_{r-1}, Q_{r-1}), i_r\big)$
$\mathsf{J}_r \downarrow \quad \uparrow \mathsf{E}_0$	
$T(n,r)$	(λ_r, P_r, Q_r)

Table B.2. The ladder.

Notations and Explanations. To define

$$\mathsf{E}_s : T(n, r-s) \times I(n, s) \longrightarrow T(n, r-s-1) \times I(n, s+1),$$

first apply E to a typical element $(\lambda_{r-s}, P_{r-s}, Q_{r-s})$ of the set $T(n, r-s)$: this gives a pair $((\lambda_{r-s-1}, P_{r-s-1}, Q_{r-s-1}), i_{r-s})$ where i_{r-s} is some element of $\underline{n}$. By definition, E_s takes the element $((\lambda_{r-s}, P_{r-s}, Q_{r-s}), i_{r-s+1} \dots i_{r-1} i_r)$ of $T(n, r-s) \times I(n, s)$ to $((\lambda_{r-s-1}, P_{r-s-1}, Q_{r-s-1}), i_{r-s} i_{r-s+1} \dots i_{r-1} i_r)$.

The map

$$\mathsf{J}_{r-s} : T(n, r-s-1) \times I(n, s+1) \longrightarrow T(n, r-s) \times I(n, s) :$$

takes (by definition)

$$((\lambda_{r-s-1}, P_{r-s-1}, Q_{r-s-1}), i_{r-s} i_{r-s+1} \dots i_{r-1} i_r) \\ \longmapsto ((\lambda_{r-s}, P_{r-s}, Q_{r-s}), i_{r-s+1} \dots i_{r-1} i_r),$$

where $(\lambda_{r-s}, P_{r-s}, Q_{r-s}) = (\lambda_{r-s-1}, P_{r-s-1}, Q_{r-s-1}) \leftarrow i_{r-s}$.

Note. To explain the top step of the ladder, take $T(n, 0)$ to be the 1-element set which contains only the triple (λ, P, Q), where $\lambda = (0, 0, \dots, 0)$ and P, Q are empty tableaux. Then identify $T(n, 0) \times I(n, r)$ with $I(n, r)$. In the same way, the bottom step is $T(n, r) \times I(n, 0) = T(n, r)$, where $I(n, 0)$ consists of the empty word only.

(B.7a) Exercise. Prove that $\mathsf{J}_{r-s} = \mathsf{E}_s^{-1}$. [Hint: use Proposition (B.6i).]

As we go up the ladder, the successive operators E_s erode $T(n, r)$, step by step, until it becomes $I(n, r)$. This progress is inverted as we go down from $I(n, r)$ to $T(n, r)$, using the operators J_s. But this "going down" is exactly described by the formula (B.2c), which means that $\mathsf{Sch} = \mathsf{J}_r \circ \mathsf{J}_{r-1} \circ \dots \circ \mathsf{J}_1$. Define

(B.7b) $\mathsf{M} := \mathsf{E}_{r-1} \circ \dots \circ \mathsf{E}_0$.

By (B.7a), M is a two-sided inverse to Sch. This proves Theorem (B.6a).

C

Schensted and Littelmann operators

C.1 Preamble

Schensted (see §§B.3, B.6) associates to every word $i \in I(n,r)$ a unique triple $(\lambda(i), P(i), Q(i)) \in T(n,r)$. This provides the following decomposition (disjoint union) of the set $I(n,r)$:

(C.1a) $$I(n,r) = \bigcup_{\lambda \in \Lambda^+(n,r)} I_\lambda(n,r),$$

where $I_\lambda(n,r)$ is the set of all $i \in I(n,r)$ such that $\lambda(i) = \lambda$, for each dominant weight $\lambda \in \Lambda^+(n,r)$. We define the *shape of a word* i to be the shape of $P(i)$ (which is also the shape of $Q(i)$). So $I_\lambda(n,r)$ is the set of all words of shape λ. In a case where n, r are supposed known, we may write $I_\lambda(n,r) = I_\lambda$.

Example. The set $I(3,3)$ is decomposed into three subsets $I_{(300)}$, $I_{(210)}$ and $I_{(111)}$; this decomposition of $I(3,3)$ is illustrated in §E.1.

Assume from now on that $\lambda \in \Lambda^+(n,r)$ is fixed.

Definition. Define two equivalence relations $\sim$ and $\approx$ on I_λ: if $i, j \in I_\lambda$ then

(C.1b) $i \sim j$ means that $P(i) = P(j)$, and

(C.1c) $i \approx j$ means that $Q(i) = Q(j)$.

We will use the following notation.

(C.1d) For any (standard) λ-tableau P whose entries are drawn from the set $\underline{n}$, let $I_\lambda(P, \sim)$ be the $\sim$ equivalence class $\{\, i \in I_\lambda \,:\, P(i) = P \,\}$, and

(C.1e) For any (standard) λ-tableau Q whose entries are $1, 2, \ldots, r$ (in some order), let $I_\lambda(Q, \approx)$ be the $\approx$ equivalence class $\{\, i \in I_\lambda \,:\, Q(i) = Q \,\}$.

Remark. If either of the tableaux P, Q is given, its shape λ is known. For this reason we will usually omit the suffix λ, and write $I_\lambda(P,\sim) = I(P,\sim)$ and $I_\lambda(Q,\approx) = I(Q,\approx)$.

The equivalence relation $\sim$ was introduced by Knuth, who proved that $\sim$ is the equivalence relation on I_λ generated by a certain collection of basic (or "elementary") moves $i \to j$, each of which affects only two places in i and j. Knuth's theorem will be proved in §§C.3, C.4. This proof is based on Knuth's paper [34, Theorem 6, p. 723].

Littelmann defines a graph G, in a wider context than here [35, p. 504]. Theorem A (see (A.4c) and Chapter D) will show that (in our present context) the equivalence relation determined by G is equal to $\approx$. We regard Theorem A as an analogue to Knuth's theorem; it says that $\approx$ is the equivalence relation on I_λ generated by a certain collection of elementary moves $i \Rightarrow j$, where $i \Rightarrow j$ means that there exists $c \in \{1, 2, \ldots, n-1\}$ such that $\tilde{f}_c(i) = j$ or such that $\tilde{e}_c(i) = j$. Notice that if $i \Rightarrow j$, then the words i and j differ in exactly one place; see (A.3g)(2).

Example. The tables in §E.1 show the $\sim$ and $\approx$ classes for the case $n = r = 3$. The $\approx$ classes are given as vertical columns in these tables; for example

$$I\left(\begin{array}{|c|c|}\hline 1 & 3 \\ \hline 2 \\ \cline{1-1}\end{array}, \approx\right) = \{211,\ 212,\ 311,\ 213,\ 312,\ 313,\ 322,\ 323\},$$

and $I\left(\begin{array}{|c|}\hline 1 \\ \hline 2 \\ \hline 3 \\ \hline\end{array}, \approx\right)$ is the one-word set $\{321\}$. The $\sim$ classes are given as horizontal rows in the tables in §E.1; for example

$$I\left(\begin{array}{|c|c|}\hline 1 & 3 \\ \hline 2 \\ \cline{1-1}\end{array}, \sim\right) = \{231,\ 213\},$$

and $I(\begin{array}{|c|c|c|}\hline 1 & 1 & 2 \\ \hline\end{array}, \sim)$ is the one-word set $\{112\}$. The one-word set $\{321\}$ is both a $\sim$ and a $\approx$ class.

C.2 Unwinding a tableau

To each tableau Y we shall associate a word KY, which may be called the *(Knuth) unwinding of* Y, as follows (see [34, p. 723] or [18, p. 17]).

Let $\lambda \in \Lambda^+(n,r)$ be a dominant weight, and let m be the number of rows of $[\lambda]$, so that $\lambda_1 \geq \lambda_2 \geq \cdots \geq \lambda_m > 0$. Define the *Knuth ordering* $<$ on $[\lambda]$ as follows (see [34, p. 723]):

(C.2a) $$\begin{aligned}(m,1) < (m,2) < \cdots &< (m,\lambda_m) \\ &< (m-1,1) < (m-1,2) < \cdots < (m-1,\lambda_{m-1}) \\ &< \cdots \\ &< (2,1) < (2,2) < \cdots < (2,\lambda_2) \\ &< (1,1) < (1,2) < \cdots < (1,\lambda_1).\end{aligned}$$

Now let $Y = (y_{a,b})_{(a,b)\in[\lambda]}$ be any λ-tableau. Define KY to be the word (C.2b) of length r obtained by writing out the entries $y_{a,b}$ according to the order (C.2a):

$$\textbf{(C.2b)}\quad KY := y_{m,1}y_{m,2}\cdots y_{m,\lambda_m}y_{m-1,1}y_{m-1,2}\cdots y_{m-1,\lambda_{m-1}} \\ \cdots\ y_{2,1}y_{2,2}\cdots y_{2,\lambda_2}y_{1,1}y_{1,2}\cdots y_{1,\lambda_1}.$$

The word KY is (by definition) the *unwinding* of the tableau Y. So KY is the word obtained by writing out the entries of each row of Y from left to right, starting with the bottom row, and working up to the first row.

Example. If $Y = \begin{array}{|c|c|c|c|}\hline 1 & 1 & 2 & 2 \\ \hline 2 & 3 \\ \cline{1-2} 3 \\ \cline{1-1}\end{array}$, then $KY = 3231122$.

Suppose that $i \in I(n,r)$. The Schensted process (see §B.3) constructs an element $(\lambda(i), P(i), Q(i))$ of $T(n,r)$, where $\lambda \in \Lambda^+(n,r)$ and $P(i)$ is a λ-tableau. Then the "unwinding" $KP(i)$ of $P(i)$ is a word, an element of $I(n,r)$. Thus we have an operation $KP : I(n,r) \to I(n,r)$, which takes each i in $I(n,r)$ to $KP(i)$. However, if we apply the Schensted process to $KP(i)$, we just get $P(i)$ again; this follows from Proposition (C.2c) below.

(C.2c) Proposition. *Let $\lambda \in \Lambda^+(n,r)$ with $\lambda_1 \geq \cdots \geq \lambda_m > 0$, and let Y be a λ-tableau.*

(i) *$P(KY) = Y$, and*

(ii) *$Q(KY)$ is completely determined by the shape λ of Y; it is the same for all λ-tableaux Y.*

The tableau $Q(KY)$ is described in (C.2h).

Proof of part (i) of Proposition (C.2c). We shall prove (i) by induction on the number m of rows of Y. If $m = 1$, then Y is a one-rowed tableau $\begin{array}{|c|c|c|c|}\hline y_{1,1} & y_{1,2} & \quad\cdots\quad & y_{1,\lambda_1} \\ \hline\end{array}$ of shape $(\lambda_1, 0, \ldots, 0)$, and $KY = y_{1,1}y_{1,2}\cdots y_{1,\lambda_1}$. We make $P(KY)$ by successively inserting $y_{1,2}, \ldots, y_{1,\lambda_1}$ into the tableau $\begin{array}{|c|}\hline y_{1,1} \\ \hline\end{array}$ (see (B.2d) and (B.3d)). But since $y_{1,1} \leq y_{1,2} \leq \cdots \leq y_{1,\lambda_1}$, each insertion simply adds a new entry to the first row. Therefore $P(KY) = \begin{array}{|c|c|c|c|}\hline y_{1,1} & y_{1,2} & \quad\cdots\quad & y_{1,\lambda_1} \\ \hline\end{array} = Y$. Thus (i) holds if $m = 1$. By the definition (B.3e), we have $Q(KY) = \begin{array}{|c|c|c|c|}\hline 1 & 2 & \cdots & \lambda_1 \\ \hline\end{array}$; this proves that (ii) also holds.

Now suppose that $m > 1$ and that Proposition (C.2c) holds for any tableau with $m-1$ rows. In particular it holds for the tableau X made by removing the first row of Y; therefore $P(KX) = X$. It is clear that $KY = KX \,|\, y_{1,1}\cdots y_{1,\lambda_1}$, hence $P(KY) = P(KX) \leftarrow y_{1,1} \leftarrow \cdots \leftarrow y_{1,\lambda_1} = X \leftarrow y_{1,1} \leftarrow \cdots \leftarrow y_{1,\lambda_1}$.

Diagram C.1 shows X, and above it, in parentheses, are the entries of row (1) of Y; these are *not* entries of X.

$$X = \begin{array}{ccccccccc}
(y_{1,1}) & (y_{1,2}) & \cdots & (y_{1,t-1}) & (y_{1,t}) & \cdots & (y_{1,\lambda_2}) & \cdots & (y_{1,\lambda_1}) \\
y_{2,1} & y_{2,2} & \cdots & y_{2,t-1} & y_{2,t} & \cdots & y_{2,\lambda_2} & \cdots & 0 \\
y_{3,1} & y_{3,2} & \cdots & y_{3,t-1} & y_{3,t} & \cdots & y_{3,\lambda_2} & \cdots & 0 \\
\vdots & \vdots & & \vdots & \vdots & & \vdots & & \\
\vdots & \vdots & & \vdots & \vdots & & y_{\beta_{\lambda_2},\lambda_2} & & \\
\vdots & \vdots & & \vdots & \vdots & & & & \\
\vdots & \vdots & & \vdots & y_{\beta_t,t} & & & & \\
\vdots & \vdots & & y_{\beta_{t-1},t-1} & 0 & \cdots & 0 & \cdots & 0 \\
\vdots & \vdots & & & & & & & \\
\vdots & y_{\beta_2,2} & & & & & & & \\
y_{\beta_1,1} & 0 & \cdots & 0 & 0 & \cdots & 0 & \cdots & 0
\end{array}$$

Diagram C.1. The tableau X, made by removing the first row from the tableau Y.

In diagram C.1, the number β_s denotes the length of column s, including the term $(y_{1,s})$. Therefore $\beta_1 = m$, and $\beta = (\beta_1, \beta_2, \ldots, \beta_{\lambda_1}, 0, \ldots, 0)$ can be regarded as the partition of r conjugate to $\lambda = (\lambda_1, \lambda_2, \ldots, \lambda_{\beta_1}, 0, \ldots, 0)$. There holds $\beta_1 \geq \beta_2 \geq \cdots \geq \beta_{t-1} \geq \beta_t \geq \cdots \geq \beta_{\lambda_1}$, but for ease of drawing, diagram C.1 illustrates a case where the β_s are distinct.

(C.2d) Definition. Let $t \in \{0, 1, 2, \ldots, \lambda_1\}$. Let $X[0] := X$, and if $t \geq 1$, define $X[t]$ to be the diagram obtained from diagram C.1 by removing the parentheses from $(y_{1,1}), (y_{1,2}), \ldots, (y_{1,t})$, and then pushing columns $(1), (2), \ldots, (t)$ down by one place.

(C.2e) Remark. For each $s \in \{1, \ldots, t\}$, column (s) of $X[t]$ is the same as column (s) of Y. In particular, $X[\lambda_1] = Y$.

(C.2f) Lemma. *Let $t \in \{0, 1, 2, \ldots, \lambda_1\}$. Then*

$$P(KX \,|\, y_{1,1} y_{1,2} \cdots y_{1,t}) = X[t].$$

In particular, $P(KY) = P(KX \,|\, y_{1,1} y_{1,2} \cdots y_{1,\lambda_1}) = X[\lambda_1] = Y$.

Thus (C.2f) will complete the proof of part (i) of (C.2c).

Proof of Lemma (C.2f). We use induction on t. If $t = 0$, the lemma claims that $X = X[0]$, which is true. So let $t \in \{1, 2, \ldots, \lambda_1\}$, and suppose that the lemma is true when t is replaced by $t-1$; that is $P(KX \,|\, y_{1,1} y_{1,2} \cdots y_{1,t-1}) = X[t-1]$. Now $P(KX \,|\, y_{1,1} y_{1,2} \cdots y_{1,t}) = P(KX \,|\, y_{1,1} y_{1,2} \cdots y_{1,t-1}) \leftarrow y_{1,t}$. So to prove Lemma (C.2f), it will be enough to prove that $X[t-1] \leftarrow y_{1,t}$ is the tableau $X[t]$ defined in (C.2d). The tableau $X[t-1]$ is displayed in

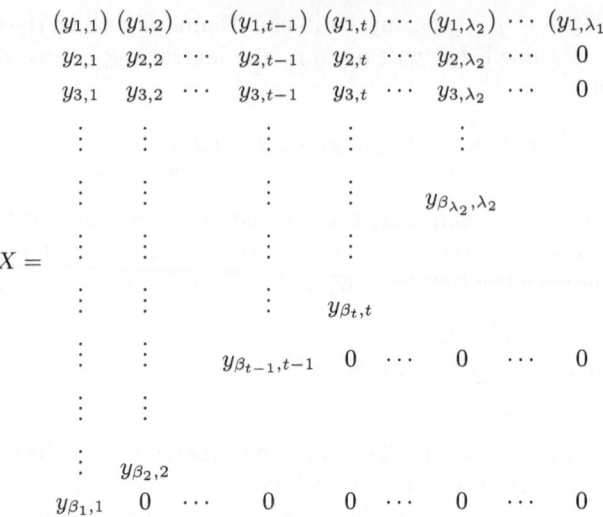

$$
X[t-1] = \begin{array}{ccccccccc}
 & & & & (y_{1,t}) & \cdots & (y_{1,\lambda_2}) & \cdots & (y_{1,\lambda_1}) \\
y_{1,1} & y_{1,2} & \cdots & y_{1,t-1} & y_{2,t} & \cdots & y_{2,\lambda_2} & \cdots & 0 \\
y_{2,1} & y_{2,2} & \cdots & y_{2,t-1} & y_{3,t} & \cdots & y_{3,\lambda_2} & \cdots & 0 \\
\vdots & \vdots & & \vdots & \vdots & & \vdots & & \\
\vdots & \vdots & & \vdots & \vdots & & y_{\beta_{\lambda_2},\lambda_2} & & \\
\vdots & \vdots & & \vdots & \vdots & & & & \\
\vdots & \vdots & & \vdots & y_{\beta_t,t} & & & & \\
\vdots & y_{\beta_2-1,2} & \cdots & y_{\beta_{t-1},t-1} & 0 & \cdots & 0 & \cdots & 0 \\
y_{\beta_1-1,1} & y_{\beta_2,2} & & & & & & & \\
y_{\beta_1,1} & 0 & \cdots & 0 & 0 & \cdots & 0 & \cdots & 0
\end{array}
$$

Diagram C.2. The tableau $X[t-1] = P(KX \mid y_{1,1}y_{1,2}\dots y_{1,t-1})$.

diagram C.2. We calculate $X[t-1] \leftarrow y_{1,t}$ by the general insertion procedure given in §B.3. To make the present notation conform with that in §B.3, take $U := X[t-1]$, $x_1 := y_{1,t}$ and μ to be the shape of U. Calculate the insertion parameters $x_1, k(1), x_2, k(2), \dots$ by the inductive rule: given the element $x_a = y_{a,t}$ for some $a \geq 1$, define $k(a)$ to be the unique element k in $\{1, \dots, \mu_a, \mu_a + 1\}$ such that

(1) $u_{a,k-1} \leq x_a < u_{a,k}$.

Then define $x_{a+1} := u_{a,k(a)}$.

It is very easy to find $k(a)$ in our case. There holds

(2) $y_{a,t-1} \leq y_{a,t} < y_{a+1,t}$,

for any $a \in \{1, \dots, \beta_t - 1\}$; the inequalities in (2) follow from the fact that Y is standard. But (2) is the same as (1), if we take $k = t$ and $x_{a+1} = y_{a+1,t}$. This shows that $k(a) = t$ and $x_{a+1} = y_{a+1,t}$. So starting with $x_1 = y_{1,t}$, which is given, we may find all insertion parameters for the insertion $X[t-1] \leftarrow y_{1,t}$. The result is given in table C.1 below. The parameter z (see the line under (B.3c)) is equal to β_t. Notice that all the $k(a)$ are equal to t, which means

a	1	2	$\cdots$	a	$\cdots$	β_t
x_a	$y_{1,t}$	$y_{2,t}$	$\cdots$	$y_{a,t}$	$\cdots$	$y_{\beta_t,t}$
$k(a)$	t	t	$\cdots$	t	$\cdots$	t

Table C.1. Insertion parameters for the insertion $X[t-1] \leftarrow y_{1,t}$.

that all the x_a lie in column t. Use (B.3d) to find $X[t-1] \leftarrow y_{1,t}$; the result is $X[t]$ as claimed in Lemma (C.2f).

Proof of part (ii) of Proposition (C.2c). We are dealing with the word $j = KY$, whose letters are indexed by the set $[\lambda]$. Fix an element $(a,s) \in [\lambda]$. Then the segment $j_{(m,1)} j_{(m,2)} \cdots j_{(a,s)}$ of j has length

(C.2g) $\psi^{(\lambda)}(a,s) := \lambda_m + \cdots + \lambda_{a+1} + s.$

This gives a bijective, order-preserving map $\psi = \psi^{(\lambda)} : [\lambda] \to \underline{r}$. We may regard $\psi = \psi^{(\lambda)}$ as the λ-tableau whose "unwinding" $K\psi$ is the word[1] $1\,2\,3 \cdots (r-1)\,r$. Notice that the tableau $\psi = \psi^{(\lambda)}$ is, in general, not standard.

Example. If $\lambda = (3,3,2,0,\ldots,0)$, then $\psi = \psi^{(\lambda)} = \begin{array}{|c|c|c|}\hline 6&7&8\\\hline 3&4&5\\\hline 1&2\\\cline{1-2}\end{array}$.

We shall prove part (ii) of Proposition (C.2c) by proving the following much stronger result:

(C.2h) Proposition (see [3, Appendix C]). *Let Y be any λ-tableau. Then*

$$Q(KY) = \mathsf{Q}^{(\lambda)},$$

where $\mathsf{Q}^{(\lambda)}$ is the λ-tableau given by $\mathsf{Q}^{(\lambda)}(a,s) := \psi^{(\lambda)}(\beta_s + 1 - a, s)$, for all $(a,s) \in [\lambda]$.

Expressed in words: $\mathsf{Q}^{(\lambda)}$ is obtained by reversing each column of the tableau $\psi^{(\lambda)}$.

Example. If $\lambda = (3,3,2,0,\ldots,0)$, then $\mathsf{Q}^{(\lambda)} = \begin{array}{|c|c|c|}\hline 1&2&5\\\hline 3&4&8\\\hline 6&7\\\cline{1-2}\end{array}$.

If Proposition (C.2h) is true, the tableau $\mathsf{Q}^{(\lambda)}$ *must* be standard, since it is the Q-symbol of a word KY (see §B.5).

Proof of Proposition (C.2h). We use induction on m. The case $m = 1$ is easy; if $Y = \begin{array}{|c|c|c|c|}\hline y_{1,1} & y_{1,2} & \cdots & y_{1,\lambda_1}\\\hline\end{array}$, then $Q(KY) = \begin{array}{|c|c|c|c|}\hline 1 & 2 & \cdots & \lambda_1\\\hline\end{array}$, which is the same as $\mathsf{Q}^{(\lambda)}$ (in this case we have $\mathsf{Q}^{(\lambda)} = \psi^{(\lambda)}$).

So now suppose that $m > 1$ and that Proposition (C.2h) is true when Y is replaced by any tableau with $m-1$ rows. In particular it is true for the tableau X obtained by removing the first row from Y, so that $Q(KX) = \mathsf{Q}^{(\lambda^*)}$, where $\lambda^* = (\lambda_2, \ldots, \lambda_m, 0, \ldots, 0) \in \Lambda^+(n, r - \lambda_1)$ is the shape of X.

[1]This word belongs to $I(r,r)$. To define $\psi^{(\lambda)}$, we should regard λ as an element of $\Lambda^+(r,r)$, but notice that $[\lambda] = [\lambda']$ if λ' is obtained from λ by adding zeros: $\lambda' = (\lambda_1, \ldots, \lambda_m, 0, 0, \ldots, 0)$.

We proved part (i) of Proposition (C.2c), namely that $P(KY) = Y$, by calculating in turn the P-symbols of the words

$$KX, \quad KX \,|\, y_{1,1}, \quad KX \,|\, y_{1,1}y_{1,2}, \quad \ldots, \quad KX \,|\, y_{1,1}y_{1,2}\cdots y_{1,\lambda_1} = KY.$$

So we shall do the same for the Q-symbols.

The first step is to find $Q(KX \,|\, y_{1,1})$. Use the procedure described in (B.3e), taking $U = X$, $x_1 = y_{1,1}$ and $r = \psi(1,1)$. (Notice that r is the length of the word $KX \,|\, y_{1,1}$). To go from $X = P(KX)$ to $P(KX \,|\, y_{1,1}) = X \leftarrow y_{1,1}$ is very easy; push down the first column of X by one place, and then put $y_{1,1}$ into the top place of that column (see proof of part (i) of Proposition (C.2c)). The tableaux X and $X \leftarrow y_{1,1}$ are shown in table C.2. To find $Q(KX \,|\, y_{1,1})$, use the recipe in (B.3e). Our induction hypothesis gives $Q(KX) = \mathsf{Q}^{(\lambda^*)}$.

The λ^*-tableau X	The $(\lambda^* + \varepsilon_1)$-tableau $X \leftarrow y_{1,1}$
$\begin{array}{ccccc} y_{2,1} & y_{2,2} & \cdots & y_{2,\lambda_2} & 0 \\ y_{3,1} & y_{3,2} & \cdots & y_{3,\lambda_2} & 0 \\ \vdots & \vdots & & \vdots & \vdots \\ \vdots & \vdots & & y_{\beta_{\lambda_2},\lambda_2} & 0 \\ \vdots & \vdots & & \vdots & \vdots \\ \vdots & y_{\beta_2,2} & \cdots & 0 & 0 \\ y_{\beta_1,1} & 0 & \cdots & 0 & 0 \end{array}$	$\begin{array}{ccccc} y_{1,1} & y_{2,2} & \cdots & y_{2,\lambda_2} & 0 \\ y_{2,1} & y_{3,2} & \cdots & y_{3,\lambda_2} & 0 \\ \vdots & \vdots & & \vdots & \vdots \\ \vdots & \vdots & & y_{\beta_{\lambda_2},\lambda_2} & 0 \\ \vdots & \vdots & & \vdots & \vdots \\ \vdots & y_{\beta_2,2} & \cdots & 0 & 0 \\ y_{\beta_1-1,1} & 0 & \cdots & 0 & 0 \\ y_{\beta_1,1} & 0 & \cdots & 0 & 0 \end{array}$

Table C.2. Inserting $y_{1,1}$ into the tableau X (P-symbol).

Diagram C.3 shows $\psi^{(\lambda^*)}$. To make $\mathsf{Q}^{(\lambda^*)}$ we reverse each column of $\psi^{(\lambda^*)}$; this

$$\psi^{(\lambda^*)} = \begin{array}{ccccc} \psi(2,1) & \psi(2,2) & \cdots & \psi(2,\lambda_2) & 0 \\ \psi(3,1) & \psi(3,2) & \cdots & \psi(3,\lambda_2) & 0 \\ \vdots & \vdots & & \vdots & \vdots \\ \vdots & \vdots & \cdots & \psi(\beta_{\lambda_2},\lambda_2) & 0 \\ \vdots & \vdots & & \vdots & 0 \\ \vdots & \psi(\beta_2,2) & & 0 & 0 \\ \psi(\beta_1,1) & 0 & \cdots & 0 & 0 \end{array} .$$

Diagram C.3. The tableau $\psi^{(\lambda^*)}$, where $\lambda^* = (\lambda_2,\ldots,\lambda_m,0,\ldots,0)$.

gives the left-hand pane in table C.3. Now follow the instructions in (B.3e): to

The Q-symbol of the word KX					The Q-symbol of the word $KX \mid y_{1,1}$				
$\psi(\beta_1, 1)$	$\psi(\beta_2, 2)$	$\cdots$	$\psi(\beta_{\lambda_2}, \lambda_2)$	0	$\psi(\beta_1, 1)$	$\psi(\beta_2, 2)$	$\cdots$	$\psi(\beta_{\lambda_2}, \lambda_2)$	0
$\vdots$	$\vdots$		$\vdots$	$\vdots$	$\vdots$	$\vdots$		$\vdots$	$\vdots$
$\psi(4, 1)$	$\psi(3, 2)$	$\cdots$	$\psi(3, \lambda_2)$	0	$\psi(4, 1)$	$\psi(3, 2)$	$\cdots$	$\psi(3, \lambda_2)$	0
$\psi(3, 1)$	$\psi(2, 2)$	$\cdots$	$\psi(2, \lambda_2)$	0	$\psi(3, 1)$	$\psi(2, 2)$	$\cdots$	$\psi(2, \lambda_2)$	0
$\psi(2, 1)$	0	$\cdots$	0	0	$\psi(2, 1)$	0	$\cdots$	0	0
					$\psi(1, 1)$	0	$\cdots$	0	0

Table C.3. Inserting $y_{1,1}$ into the tableau X (Q-symbol).

go from the left-hand pane to the right-hand pane, we adjoin a new place—this must be the place which is new when we go from $P(KX)$ to $P(KX) \leftarrow y_{1,1}$, namely the place $(\beta_1, 1)$. And in this place we must put "r", which in our case is $\psi(1, 1)$. This gives $Q(KX \mid y_{1.1})$ shown in the right-hand pane of table C.3.

We go on to insert $y_{1,2}, \ldots, y_{1,\lambda_1}$ in turn. At each insertion, say $y_{1,t}$, we adjoin $\psi(1, t)$ to the bottom of column (t). But the new column (t) so made is the same as column (t) of $\mathsf{Q}^{(\lambda)}$. When we have inserted y_{1,λ_1} we have the complete tableau $\mathsf{Q}^{(\lambda)}$. This finishes the proof of Proposition (C.2h).

Hence we have proved Proposition (C.2c).

(C.2i) Exercise. Let $\lambda \in \Lambda^+(n, r)$, and let i be any element of $I_\lambda(n, r)$. Then $i = KP(i)$ if and only if $Q(i) = \mathsf{Q}^{(\lambda)}$. In other words, the $\approx$ class of $\mathsf{Q}^{(\lambda)}$ consists of all $i \in I_\lambda(n, r)$ which satisfy $i = KP(i)$.

[Hint. Let $i \in I(n, r)$. Schensted's Theorem (B.6a) tells us that

(i) $i = KP(i)$ if and only if $\mathsf{Sch}(i) = \mathsf{Sch}(KP(i))$.

Now assume that $i \in I_\lambda(n, r)$, which means that $\lambda(i) = \lambda$ (see (C.1a)). Hence

(ii) $\mathsf{Sch}(i) = (\lambda, P(i), Q(i))$.

To calculate $\mathsf{Sch}(KP(i))$ we take $Y = P(i)$ in (C.2c). This shows that $P(KP(i)) = P(i)$. Since $P(i)$ has shape λ, it follows that $\lambda(KP(i)) = \lambda$. But (C.2c)(ii) and (C.2h) tell us that $Q(KP(i)) = \mathsf{Q}^{(\lambda)}$. Therefore

(iii) $\mathsf{Sch}(KP(i)) = (\lambda, P(i), \mathsf{Q}^{(\lambda)})$.

Now (i), together with (ii) and (iii), give the desired result: $i = KP(i)$ if and only if $Q(i) = \mathsf{Q}^{(\lambda)}$.]

C.3 Knuth's theorem

(C.3a) Theorem (see [34, p. 723]). *Let i, j be words in $I(n,r)$. Then $i \sim j$ (i.e. $P(i) = P(j)$) if and only if there is a finite sequence of words*

(C.3b) $i(1), i(2), \ldots, i(s)$

such that $i(1) = i$, $i(s) = j$ and each consecutive pair of words $i(\sigma - 1)$, $i(\sigma)$ is connected by a basic *(or* elementary*) move of type K' or K''.*

These basic moves are as follows [34, p. 723].

Definition. A move of type K' changes a word

(C.3c) $\ldots bca \ldots$ to $\ldots bac \ldots$,

where a, b, c are letters (i.e. elements of $\underline{n}$) such that $a < b \leq c$.
A move of type K'' changes a word

(C.3d) $\ldots acb \ldots$ to $\ldots cab \ldots$,

where a, b, c are letters (i.e. elements of $\underline{n}$) such that $a \leq b < c$.

Remarks.

(i) Each basic move is assumed to be *symmetric*, i.e. if a move takes a word w to another word w', then it also takes w' to w.
(ii) In (C.3c) and (C.3d), the symbol $\ldots$ stands for a word (possibly empty) which is not changed in the move. For example the type K' move (C.3c) changes $BbcaC$ to $BbacC$, where B, C are fixed words. (By (i), this move also takes $BbacC$ to $BbcaC$.)

The "only if" part of Knuth's theorem will be proved in this section, and the "if" part will be proved in §C.4. So in this section (§C.3) we must prove

(C.3e) *If $i, j \in I(n,r)$ are such that $P(i) = P(j)$, then i can be connected to j by a finite sequence of basic moves.*

The essence of this is that *every insertion operation U to $U \leftarrow x$ can be broken down into a sequence of basic moves.* The next proposition puts this fact in a form suitable for our purposes; in (C.3p) it will be shown that (C.3f) implies (C.3e).

(C.3f) Proposition. *Let $r > 1$, $\mu \in \Lambda^+(n, r-1)$. Let U be any μ-tableau and x any element of $\underline{n}$. Regard KU and x as words of lengths $r-1$ and 1 respectively, so that the "concatenation" $w = KU \mid x$ of these may be regarded as an element in $I(n,r)$. Then there is a finite sequence of basic moves in $I(n,r)$ which takes w to the word $K(U \leftarrow x)$.*

Proof. We shall give in (C.3i)–(C.3k) an explicit sequence of basic moves which takes w to $K(U \leftarrow x)$[2].

It is desirable to fix first some notation for the words which will be used in the proof of (C.3f).

(C.3g) Notation for words and places. All the words in this section have length r, and their entries are labelled by the set of places $[\mu] \cup \{(r)\}$. The $r-1$ elements of $[\mu]$ are arranged according to the Knuth order (see (C.2a)), and (r) is the last place. Therefore if $\mu = (\mu_1, \dots, \mu_m, 0, \dots, 0) \in \Lambda^+(n, r-1)$ (with $\mu_1 \geq \dots \geq \mu_m > 0$ and $\sum \mu_j = r - 1$), then a typical word looks like this: $y = y_{m,1} \cdots y_{m,\mu_m} \;\cdots\; y_{1,1} \cdots y_{1,\mu_1} y_r$.

To resume the proof of (C.3f), write $x = x_1$ and $w = KU \mid x$, so that $w = u_{m,1} \dots u_{m,\mu_m} \;\cdots\; u_{1,1} \dots u_{1,\mu_1} x_1$. Recall from (B.3b) the "parameters" of the insertion of x_1 into the tableau U: for each $a \in \{1, \dots, z\}$, define $x_a := u_{a-1,k(a-1)}$ if $a > 1$, or if $a = 1$ define $x_1 = x$. Define $k(a)$ to be the smallest $k \in \{1, 2, \dots, \mu_a, \mu_a + 1\}$ such that $x_a < u_{a,k}$. If $k(a) = \mu_a + 1$ (which means that $x_a \geq u_{a,\mu_a}$), then the insertion sequence stops at this stage. Define z to be the first a such that $x_a \geq u_{a,\mu_a}$. The tableau $U \leftarrow x_1$ is denoted $P = (p_{a,b})_{(a,b) \in [\lambda]}$, where $\lambda = \mu + \varepsilon_z$ (see (B.3d)).

According to (B.3d), each row $a > z$ of the tableau U, is identical to the corresponding row of $P = U \leftarrow x_1$; and (B.3d)(1°) shows that also $u_{z,t} = p_{z,t}$ for all $t \in \{1, \dots, \mu_z\}$. Therefore

(C.3h) $u_{a,t} = p_{a,t}$ for all places $(a,t) \leq (z, \mu_z)$.

Next define a sequence of words $\xi(a,t)$, one for each place $(a,t) \in [\mu]$, which will "interpolate" between the words $w = KU \mid x_1$ and $KP = K(U \leftarrow x_1)$. Use the following notation: if $\tau \in [\mu]$, then $\tau+$ (respectively $\tau-$) denotes the place immediately after (respectively immediately before) τ in the order (C.3g) of $[\mu] \cup \{(r)\}$. For example, if $a \in \{2, \dots, z\}$, then $(a,t)+$ is $(a, t+1)$ for all $1 \leq t < \mu_a$, and $(a, \mu_a)+ = (a-1, 1)$.

Definition of the words $\xi(a,t)$.

(C.3i) If $(a,t) \leq (z, \mu_z)$, then define $\xi(a,t) := KP$.

(C.3j) If $(a,t) > (z, \mu_z)$ and $k(a)+1 \leq t \leq \mu_a$, then define $\xi(a,t)_{(a,t)+} := x_a$, $\xi(a,t)_\tau := u_\tau$ if $\tau \leq (a,t)$, and $\xi(a,t)_\tau := p_{\tau-}$ if $\tau > (a,t)+$.

(C.3k) If $(a,t) > (z, \mu_z)$ and $1 \leq t \leq k(a)$, then define $\xi(a,t)_{(a,t)} := x_{a+1}$, $\xi(a,t)_\tau := u_\tau$ if $\tau < (a,t)$, and $\xi(a,t)_\tau := p_{\tau-}$ if $\tau > (a,t)$.

---

[2]In fact the construction of these basic moves is an essential part of Knuth's proof of his theorem; see [34, end of p. 723, and first 7 lines of p. 724].

(C.3l) Pivot of $\xi(a,t)$. Assume that $(a,t) > (z,\mu_z)$. Then the word $\xi(a,t)$ can be described as follows: define the *pivot of* $\xi(a,t)$ to be the pair $(x_a,(a,t)+)$ in case (C.3j), and to be $(x_{a+1},(a,t))$ in case (C.3k). Then in both cases the rule is: at every place τ left of the pivot, let $\xi(a,t)_\tau = u_\tau$, and at every place τ right of the pivot, let $\xi(a,t)_\tau = p_{\tau-}$. At the pivot itself, $\xi(a,t)$ has entry x_a in case (C.3j), or x_{a+1} in case (C.3k). If a word is among the $\xi(a,t)$, it is completely determined by its pivot.

(C.3m) Proposition.

(i) $\xi(1,\mu_1) = w = KU \mid x_1$.

(ii) $\xi(z-1,1) = KP = K(U \leftarrow x_1)$.

(iii) $\xi(a,1) = \xi(a+1,\mu_{a+1})$, *for all* $a \in \{1,\ldots,m-1\}$.

(iv) *If* $k(a)+1 \le t \le \mu_a$, *there is a basic move of type* K' *which takes* $\xi(a,t)$ *to* $\xi(a,t-1)$.

(v) *If* $2 \le t \le k(a)$, *there is a basic move of type* K'' *which takes* $\xi(a,t)$ *to* $\xi(a,t-1)$.

Proof. (i) The pivot of $\xi(1,\mu_1)$ is $(x_1,(r))$, therefore $\xi(1,\mu_1)$ has u_τ in each place $\tau \in [\mu]$, and x_1 in place (r). Hence $\xi(1,\mu_1) = KU \mid x_1$.

(ii) The pivot of $\xi(z-1,1)$ is $(x_z,(z-1,1))$ since $1 \le t \le k(z-1)$ for $t=1$. Thus $\xi(z-1,1)$ has x_z at place $(z-1,1)$. At a place $\tau < (z-1,1)$, the entry in $\xi(z-1,1)$ is u_τ, and this equals p_τ, by (C.3h). At $t > (z-1,1)$, the entry is $p_{\tau-}$. Therefore $\xi(z-1,1) = KP$.

(iii) All that is needed, is to show that both $\xi(a,1)$ and $\xi(a+1,\mu_{a+1})$ have the same pivot $(x_{a+1},(a,1))$. We leave this as an exercise for the reader.

(iv) Suppose that $k(a)+1 \le t \le \mu_a$. By definition (C.3j), the entries of $\xi(a,t)$ at the places $(a,t)-$, (a,t), $(a,t)+$ are $u_{(a,t)-}$, $u_{a,t}$, x_a, respectively. If these entries are denoted b, c, a, then we shall prove that $a < b \le c$. The inequality $b \le c$, i.e. $u_{(a,t)-} \le u_{(a,t)}$, follows from $(a,t)- = (a,t-1)$ and the standardness of U (note that $k(a)+1 \le t$ implies that $2 \le t$). To see that $a < b$, use the inequality $u_{a,k(a)-1} \le x_a < u_{a,k(a)}$ (see (B.3c)). Standardness of U gives $u_{a,k(a)} \le u_{a,t-1}$, since $k(a)+1 \le t$. Therefore $x_a < u_{a,k(a)} \le u_{a,t}$, hence $a < b$.

We may now make a move of type K' which interchanges a and c, and leaves $\xi(a,t)$ otherwise unchanged. At the places (a,t), $(a,t)+$, the word $K'\xi(a,t)$ has entries x_a, $u_{a,t}$. However $u_{a,t} = p_{a,t}$ since $t \ne k(a)$. It is now easy to see that $K'\xi(a,t) = \xi(a,t-1)$.

(v) The proof is on the same lines as that of (iv). Suppose $2 \le t \le k(a)$. By (C.3k) the entries in $\xi(a,t)$ at the places $(a,t)- = (a,t-1)$, (a,t), $(a,t)+$ are $u_{a,t-1}$, x_{a+1}, $p_{a,t}$, respectively. If these entries are denoted a, c, b, we leave it to the reader to prove that $a \le b < c$. This shows that there is a move of type K'' which interchanges a, c, and leaves $\xi(a,t)$ otherwise unchanged. Therefore the entries in $K''\xi(a,t)$ at places $(a,t-1)$, (a,t), are x_{a+1}, $u_{a,t-1}$. But $u_{a,t-1} = p_{a,t-1}$ because $t-1 \ne k(a)$. It follows that $K''\xi(a,t) = \xi(a,t-1)$.

This completes the proof of Proposition (C.3m).

And this proves Proposition (C.3f).

(C.3n) Example. Take $\mu = (4, 2, 1, 1) \in \Lambda^+(4, 8)$, and U as given in (C.3o). Then U is a μ-tableau. Now take $x_1 = 1$. We calculate $P = U \leftarrow x_1$ by the methods of §B.4. This tableau also is given in (C.3o). It is a λ-tableau, where $\lambda = (4, 2, 2, 1) \in \Lambda^+(4, 9)$.

(C.3o)
$$U = \begin{array}{|c|c|c|c|} \hline 1 & 1 & 2 & 2 \\ \hline 2 & 4 \\ \cline{1-2} 3 \\ \cline{1-1} 4 \\ \cline{1-1} \end{array}\,, \qquad P = \begin{array}{|c|c|c|c|} \hline 1 & 1 & 1 & 2 \\ \hline 2 & 2 \\ \cline{1-2} 3 & 4 \\ \cline{1-2} 4 \\ \cline{1-1} \end{array}\,.$$

The parameters for the insertion of $x_1 = 1$ are as follows (see (B.3b)): $z = 3$ and

$$x_1 = 1\ (= p_{1,k(1)}),\ x_2 = 2\ (= u_{1,k(1)} = p_{2,k(2)}),\ x_3 = 4\ (= u_{2,k(2)} = p_{3,k(3)}),$$
$$k(1) = 3, \qquad k(2) = 2, \qquad k(3) = 2.$$

It is rather easy to display the words $\xi(a, t)$, for all $(a, t) \in [\mu]$ (see table C.4). By (C.3m)(i),(ii) we know that $\xi(1, 4) = KU \mid x_1$, and $\xi(2, 1) = KP$; so write in these words. If $(a, t) \leq (z, \mu_z)$ then $\xi(a, t) = KP$ by (C.3i), and this gives us $\xi(3, 1)$ and $\xi(4, 1)$. If $(a, t) > (z, \mu_z)$, determine the pivot of $\xi(a, t)$, using (C.3j) or (C.3k) as appropriate. For example the pivot of $\xi(1, 4)$ is $(x_1, (1, 4)+) = (x_1, (9))$, and the pivot of $\xi(1, 2)$ is $(x_2, (1, 2))$. For each $\xi(a, t)$, we have underlined the first term (x_a or x_{a+1}) in the pivot of $\xi(a, t)$ in table C.4.

Proof of the "only if" part of Knuth's theorem. Suppose $i \in I(n, r)$, and for each $s \in \{0, 1, \ldots, r\}$ define $P_s(i) = P(i_1 \ldots i_s)$ (take $P_0(i)$ to be the empty tableau). Let $s \in \{1, 2, \ldots, r\}$ and let $U = P_{s-1}(i)$ and $x = i_s$. Then Proposition (C.3f) provides a sequence of words $\xi(a, t)$, and hence a sequence of basic moves taking the word $KP_{s-1}(i) \,|\, i_s$ to the word $K(P_{s-1}(i) \leftarrow i_s) = KP_s(i)$. Using the notation of §B.7 (the "ladder"), we may now construct a sequence of basic moves taking $KP_{s-1} \,|\, i_s i_{s+1} \ldots i_r$ to $KP_s \,|\, i_{s+1} \ldots i_r$; we simply use the sequence of words $\xi(a, t) \,|\, i_{s+1} \ldots i_r$ in place of the $\xi(a, t)$. Since we can do this for each $s = 1, 2, \ldots, r$, we deduce the following fundamental proposition:

(C.3p) Proposition. *Given $i \in I(n, r)$, there is a sequence of basic moves in $I(n, r)$ which takes i to $KP(i)$.*

It is now easy to prove (C.3e), which is the "only if" part of Knuth's theorem (C.3a). For given $i, j \in I(n, r)$ such that $P(i) = P(j)$, we use (C.3p) to make two sequences of basic moves, one taking i to $KP(i)$ and one taking j to $KP(j) = KP(i)$. Then the first of these sequences, followed by the "reverse" of the second, takes i to j.

Place	$(4,1)$	$(3,1)$	$(2,1)$	$(2,2)$	$(1,1)$	$(1,2)$	$(1,3)$	$(1,4)$	(9)	Move
$\xi(1,4)$	$u_{4,1}$	$u_{3,1}$	$u_{2,1}$	$u_{2,2}$	$u_{1,1}$	$u_{1,2}$	$u_{1,3}$	$u_{1,4}$	$\underline{x_1}$	K'
$\xi(1,3)$	$u_{4,1}$	$u_{3,1}$	$u_{2,1}$	$u_{2,2}$	$u_{1,1}$	$u_{1,2}$	$\underline{x_2}$	x_1 $=p_{1,3}$	$p_{1,4}$ $=u_{1,4}$	K''
$\xi(1,2)$	$u_{4,1}$	$u_{3,1}$	$u_{2,1}$	$u_{2,2}$	$u_{1,1}$	$\underline{x_2}$	$p_{1,2}$	$p_{1,3}$	$p_{1,4}$	K''
$\xi(1,1)$	$u_{4,1}$	$u_{3,1}$	$u_{2,1}$	$u_{2,2}$	$\underline{x_2}$	$p_{1,1}$ $=u_{1,1}$	$p_{1,2}$	$p_{1,3}$	$p_{1,4}$	–
$\xi(2,2)$	$u_{4,1}$	$u_{3,1}$	$u_{2,1}$	$\underline{x_3}$	x_2 $=p_{2,2}$	$p_{1,1}$	$p_{1,2}$	$p_{1,3}$	$p_{1,4}$	K''
$\xi(2,1)$	$p_{4,1}$ $=u_{4,1}$	$p_{3,1}$ $=u_{3,1}$	$\underline{x_3}$ $=p_{3,2}$	$p_{2,1}$	$p_{2,2}$	$p_{1,1}$	$p_{1,2}$	$p_{1,3}$	$p_{1,4}$	–
$\xi(3,1)$	$p_{4,1}$	$p_{3,1}$	$p_{3,2}$	$p_{2,1}$	$p_{2,2}$	$p_{1,1}$	$p_{1,2}$	$p_{1,3}$	$p_{1,4}$	–
$\xi(4,1)$	$p_{4,1}$	$p_{3,1}$	$p_{3,2}$	$p_{2,1}$	$p_{2,2}$	$p_{1,1}$	$p_{1,2}$	$p_{1,3}$	$p_{1,4}$	–

Table C.4. The sequence of words associated to Schensted insertion.

C.4 The "if" part of Knuth's theorem

Let n, r be positive integers. In this section we will prove the "if" part of Knuth's theorem (C.3a), that is: if $i, j \in I(n,r)$ can be connected by a sequence of basic moves as in (C.3b), then $i \sim j$ (i.e. $P(i) = P(j)$).

Clearly it will be enough to prove:

(C.4a) If i and j in $I(n,r)$ are connected by a basic move, then $P(i) = P(j)$.

For a standard tableau U and letters $x_1, \ldots, x_k$, we write

$$U \leftarrow x_1 x_2 \cdots x_k = (\cdots((U \leftarrow x_1) \leftarrow x_2) \cdots) \leftarrow x_k$$

to ease the reading. We will prove the following

(C.4b) Proposition (see [34, pp. 721, 722]). *Let U be a standard tableau with entries drawn from $\underline{n}$, and let $a, b, c \in \underline{n}$.*

(1) *If $a < b \leq c$, then $U \leftarrow bac = U \leftarrow bca$.*

(2) *If $a \leq b < c$, then $U \leftarrow acb = U \leftarrow cab$.*

This proposition implies (C.4a), by means of a simple induction on the length r of i and j.

Our proof of the proposition builds on Schensted's original description of the insertion process $U \leftarrow x$, which reads as follows. Let U be a μ-tableau and x be a letter.

If U is the empty tableau or $u_{1,\mu_1} \leq x$ (so that the insertion sequence (B.3b) has length $z = 1$), then $U \leftarrow x$ is obtained from U by appending the letter x to the first row of U:

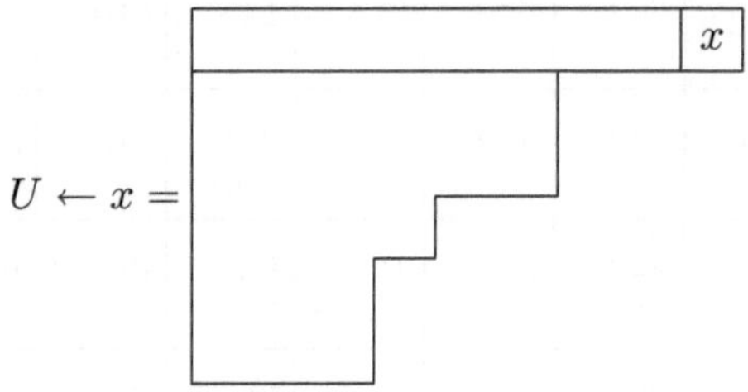

If U is not empty and $x < u_{1,\mu_1}$ (so that the insertion sequence (B.3b) has length $z > 1$), then choose $k \leq \mu_1$ minimal with $x < u_{1,k}$ and set $y = u_{1,k}$. The tableau $U \leftarrow x$ has first row $(u_{1,1}, \ldots, x, \ldots, u_{1,\mu_1})$ (with x in column k), while the remaining rows of $U \leftarrow x$ are given by $\tilde{U} \leftarrow y$. Here $\tilde{U}$ is the "sub-tableau" of U obtained from U by removing the first row. In illustrative terms:

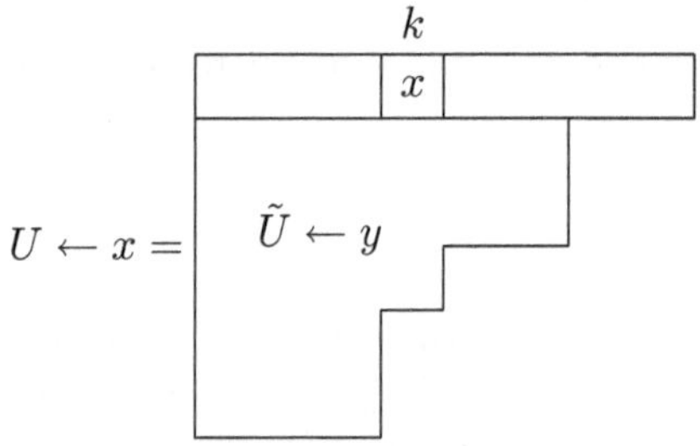

Of course, this description of $U \leftarrow x$ follows directly from our description (B.3d).

The idea of proof for the proposition is this. In either case (1) or (2), check that the first rows of the two tableaux shown coincide. Then consider the tableaux obtained by removing the first rows.

If any of the three letters a, b, c does not bump a letter into the second row, then it is fairly easy to see that these "sub-tableaux" are equal. If all three letters a, b, c bump letters into the second row—x, y, z, say—then the letters x, y, z can be shown to satisfy either (1) or (2). We can then conclude by induction on the number of rows of U.

Before we give the proof of (C.4b), let's look at two examples.

Example 1. Let $n = 5$, $a = 1$, $b = 1$ and $c = 3$ (so that part (2) of the proposition applies), and consider the tableau

$$U = \begin{array}{|c|c|c|c|c|c|}\hline 1&1&2&3&3&4\\\hline 2&2&3&4\\\cline{1-4} 3&4&4&5\\\cline{1-4}\end{array}$$

We have

$$\tilde{U} = \begin{array}{|c|c|c|c|}\hline 2&2&3&4\\\hline 3&4&4&5\\\hline\end{array}\,.$$

Let us concentrate on the first rows of $U \leftarrow acb$ and $U \leftarrow cab$. We get

$$U \leftarrow acb = U \leftarrow 131 = \begin{array}{|c|c|c|c|c|c|}\hline 1&1&1&1&3&3\\\hline \multicolumn{4}{|c|}{\tilde{U} \leftarrow 243}\\\cline{1-4}\end{array}$$

from Schensted's inductive description, because $a = 1$ bumps $x = 2$ from the first row of U, $c = 3$ bumps $z = 4$ from the first row of $U \leftarrow a$, and $b = 1$ bumps $y = 3$ from the first row of $U \leftarrow ac$.

Similarly, we get

$$U \leftarrow cab = U \leftarrow 311 = \begin{array}{|c|c|c|c|c|c|}\hline 1&1&1&1&3&3\\\hline \multicolumn{4}{|c|}{\tilde{U} \leftarrow 423}\\\cline{1-4}\end{array}$$

Applying (C.4b)(2), it follows that $\tilde{U} \leftarrow 243 = \tilde{U} \leftarrow 423$ (by induction on the number of rows), hence also $U \leftarrow 131 = U \leftarrow 311$.

Example 2. Let $n = 5$ again, $a = 1$, $b = 1$ and $c = 2$ (so that part (2) of proposition (C.4b) applies again), and consider the tableau

$$U = \begin{array}{|c|c|c|c|c|c|}\hline 1&1&3&3&3&3\\\hline 2&3&4&4&4\\\cline{1-5} 4&5&5&5\\\cline{1-4}\end{array}$$

We have

$$\tilde{U} = \begin{array}{|c|c|c|c|c|}\hline 2&3&4&4&4\\\hline 4&5&5&5\\\cline{1-4}\end{array}\,.$$

In this case, we get

$$U \leftarrow acb = U \leftarrow 121 = \begin{array}{|c|c|c|c|c|c|}\hline 1&1&1&1&3&3\\\hline \multicolumn{5}{|c|}{\tilde{U} \leftarrow 332}\\\cline{1-5}\end{array}$$

from Schensted's inductive description, because $a = 1$ bumps $x = 3$ from the first row of U, $c = 2$ bumps $z = 3$ from the first row of $U \leftarrow a$, and $b = 1$ bumps $y = c = 2$ from the first row of $U \leftarrow ac$. Similarly, we get

$$U \leftarrow cab = U \leftarrow 211 = \begin{array}{|c|c|c|c|c|c|}\hline 1&1&1&1&3&3\\\hline \multicolumn{5}{|c|}{\tilde{U} \leftarrow 323}\\\cline{1-5}\end{array}$$

This time applying (C.4b)(1), it follows that $\tilde{U} \leftarrow 332 = \tilde{U} \leftarrow 323$ (by induction on the number of rows), hence also $U \leftarrow 121 = U \leftarrow 211$.

Proof of Proposition (C.4b). The proof is done by induction on the number m of rows of U.

If $m = 0$, then U is the empty tableau and

$$(1) \quad U \leftarrow bac = \begin{array}{|c|c|}\hline a & c \\ \hline b & \\ \cline{1-1}\end{array} = U \leftarrow bca \text{ if } a < b \leq c,$$

$$(2) \quad U \leftarrow acb = \begin{array}{|c|c|}\hline a & b \\ \hline c & \\ \cline{1-1}\end{array} = U \leftarrow cab \text{ if } a \leq b < c.$$

Thus (C.4b) holds in case $m = 0$.

Suppose $m > 0$. Let μ denote the shape of U and write $U = (u_{x,y})_{(x,y)\in[\mu]}$. Furthermore, let $\tilde{U}$ be the tableau obtained from U by removing the first row.

We shall prove the parts (1) and (2) separately.

Proof of part (1) of Proposition (C.4b). Assume we have $a, b, c \in \underline{n}$ such that $a < b \leq c$. We want to prove that $U \leftarrow bac = U \leftarrow bca$.

To find the tableau $W := U \leftarrow b$, let b bump $u_{1,l}$ into row (2), where l is the smallest element of $\{1, \ldots, \mu_1, \mu_1 + 1\}$ such that $b < u_{1,l}$. This means:

(i) row (1) of W is the same as row (1) of U except at place $(1, l)$, where $w_{1,l} = b$, and

(ii) the tableau obtained by removing the first row of W, is $\tilde{W} = \tilde{U} \leftarrow y$, where $y := u_{1,l}$. (If $l = \mu_1 + 1$, so that $y = \infty$, we make the convention that inserting ∞ into row (2) has no effect on $\tilde{U}$.)

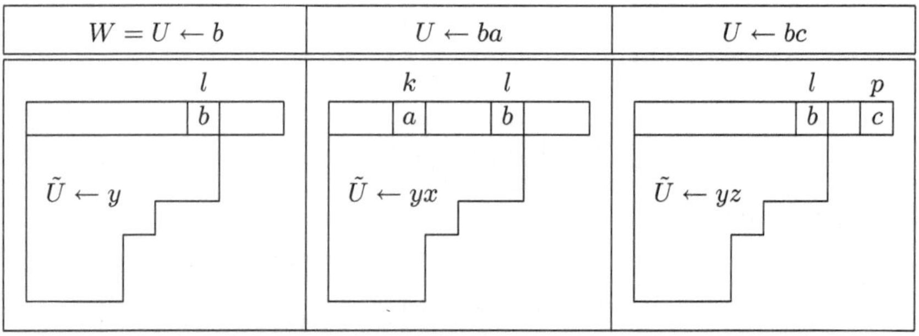

Table C.5. Inserting b, a and b, c into U when $a < b \leq c$.

To find the tableaux $U \leftarrow ba = W \leftarrow a$ and $U \leftarrow bc = W \leftarrow c$, define k to be the smallest element of $\{1, \ldots, \mu_1, \mu_1 + 1\}$ and p to be the smallest element of $\{1, \ldots, \mu_1 + 1, \mu_1 + 2\}$ such that $a < w_{1,k}$ and $c < w_{1,p}$, respectively. (The case $p = \mu_1 + 2$ only occurs when $k = \mu_1 + 1$.)

Then

(iii) $k \leq l$ (because $a < b = w_{1,l}$), and

(iv) $l < p$ (because $w_{1,l} = b \le c < w_{1,p}$).

Make $U \leftarrow ba$ from W by letting a bump $x := w_{1,k}$ into row (2); make $U \leftarrow bc$ from W by letting c bump $z := w_{1,p}$ into row (2). The resulting tableaux are shown in table C.5.

To find $U \leftarrow bac$, insert c into the tableau $W' := U \leftarrow ba$. First find the smallest p' in $\{1, \ldots, \mu_1, \mu_1 + 1, \mu_1 + 2\}$ such that $c < w'_{1,p'}$. But any p' such that $c < w'_{1,p'}$ is $> l$ (because $w'_{1,l} = b \le c$), and all the entries $w'_{1,s}$ in row (1) of W' such that $s \ge l$, coincide with the corresponding entries $w_{1,s}$ in row (1) of W, because the process which takes W to $W' = W \leftarrow a$ affects only the part of row (1) to the left of $(1, l)$. Therefore $p = p'$, and $U \leftarrow bac$ is shown in the left pane of table C.6. An entirely similar argument gives $U \leftarrow bca$, using the fact that the process which takes W to $W'' = W \leftarrow b$ affects only the part of row (1) to the right of $(1, l)$; this tableau is shown in the right pane of table C.6. We next prove that $x < y \le z$. First, to prove $x < y$, observe

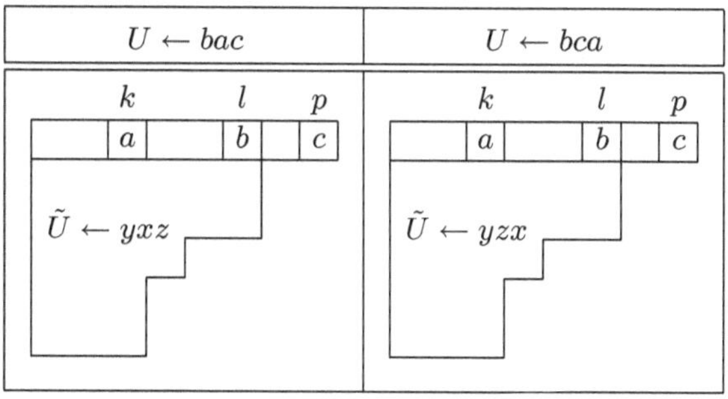

Table C.6. Inserting b, a, c and b, c, a into U when $a < b \le c$.

that (iii) gives $x = w_{1,k} \le w_{1,l} = b < u_{1,l} = y$. Second, to prove $y \le z$, observe that (iv) gives $y = u_{1,l} \le u_{1,p} = z$. This argument is also valid when $y = \infty$, because then $z = \infty$ as well.

It follows that $\tilde{U} \leftarrow yxz = \tilde{U} \leftarrow yzx$ by the induction hypothesis; and table C.6 gives the desired result $U \leftarrow bac = U \leftarrow bca$.

There is still one "loose end" to be tidied up! Namely it can happen that $k = l$; the argument above still works, but we must re-draw the first rows of the tableaux shown in table C.6. Each of these rows (which are equal) looks like

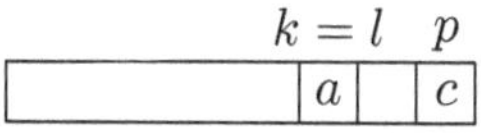

This concludes the proof of Proposition (C.4b)(1).

Proof of part (2) of Proposition (C.4b). We now assume that $a \le b < c$ and show that $U \leftarrow acb = U \leftarrow cab$.

To find the tableaux $U \leftarrow a$ and $U \leftarrow c$, define k and p to be the smallest elements of $\{1, \ldots, \mu_1, \mu_1 + 1\}$ such that $a < u_{1,k}$ and $c < u_{1,p}$, respectively.

As usual, make $V := U \leftarrow a$ from U by letting a bump $x := u_{1,k}$ into row (2), and make $W := U \leftarrow c$ from U by letting c bump $z := u_{1,p}$ into row (2). Then $k \leq p$ (because $a < c$). We consider two cases:

Case 1. $k < p$. Then $k \leq \mu_1$, hence the first row of V has μ_1 entries. To find the tableau $U \leftarrow ac = V \leftarrow c$ define p' to be the smallest element of $\{1, \ldots, \mu_1, \mu_1 + 1\}$ such that $c < v_{1,p'}$. But any p' with $c < v_{1,p'}$ is $> k$ (because $v_{1,k} = a < c$), and all the entries $v_{1,s}$ in row (1) of V with $s > k$, coincide with the corresponding entries $u_{1,s}$ of U (because the first rows of U and V differ only at place $(1, k)$). It follows that $p = p'$, and c bumps the letter $z = u_{1,p}$ of V into row (2). An entirely similar argument shows that a bumps the letter $x = u_{1,k} = w_{1,k}$ of W into row (2).

The tableaux $V \leftarrow c$ and $W \leftarrow a$ are shown in table C.7.

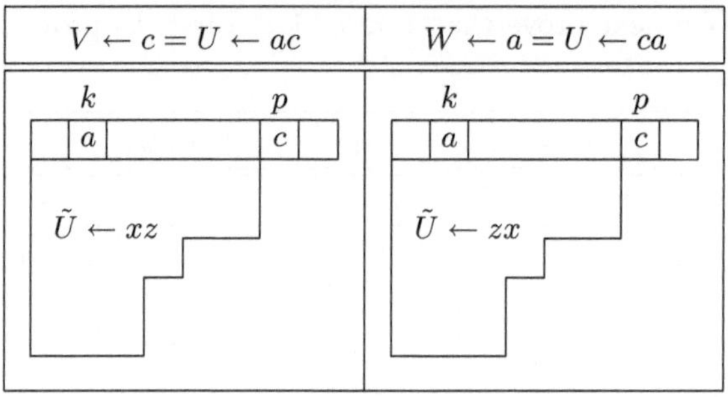

Table C.7. Inserting a, c and c, a into U when $a \leq b < c$ and $k < p$.

The first rows of $V \leftarrow c$ and $W \leftarrow a$ coincide. Hence b bumps the same letter $y = v_{1,l} = w_{1,l}$ into row (2) of $V \leftarrow c$ and $W \leftarrow a$. Furthermore, it follows from $a \leq b < c$ that $k < l \leq p$. The tableaux $U \leftarrow acb = V \leftarrow cb$ and $U \leftarrow cab = W \leftarrow ab$ are displayed in table C.8.

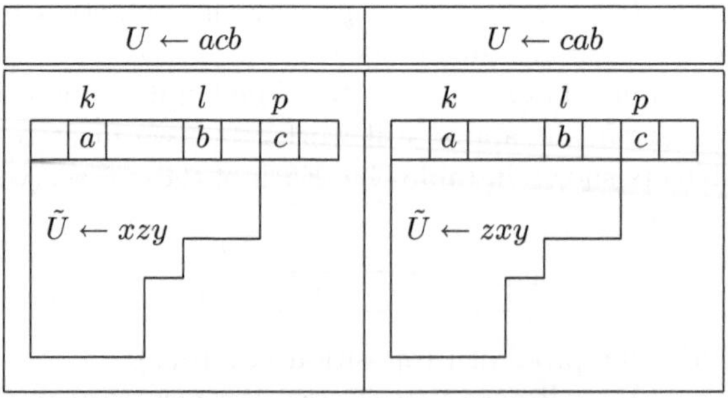

Table C.8. Inserting a, c, b and c, a, b into U when $a \leq b < c$ and $k < p$.

We next prove that $x \le y < z$. First, we have $x = u_{1,k} \le u_{1,l} = y$ if $l < p$, and $x = v_{1,k} \le v_{1,p-1} \le c = y$ if $l = p$. Second, $y = v_{1,l} \le v_{1,p} = c < u_{1,p} = z$.

It follows that $\tilde{U} \leftarrow xzy = \tilde{U} \leftarrow zxy$ by the induction hypothesis; and table C.8 gives $U \leftarrow acb = U \leftarrow cab$ in case $k < p$.

Case 2. $k = p$. Then $x = u_{1,k} = u_{1,p} = z$. To find the tableau $V \leftarrow c$ in this case, define p' to be the smallest element of $\{1, \ldots, \mu_1 + 1, \mu_1 + 2\}$ such that $c < v_{1,p'}$. But $v_{1,k} = a < c < u_{1,k} \le u_{1,k+1} = v_{1,k+1}$, hence $p' = k + 1$. Therefore c bumps the letter $w = v_{1,k+1} = u_{1,k+1}$ of V into row (2).

To find the tableau $W \leftarrow a$, note that $w_{1,p-1} = u_{1,p-1} \le a < c = w_{1,p}$. Therefore a bumps the letter $c = w_{1,p}$ of W into row (2).

The tableaux $V \leftarrow c$ and $W \leftarrow a$ are shown in table C.9.

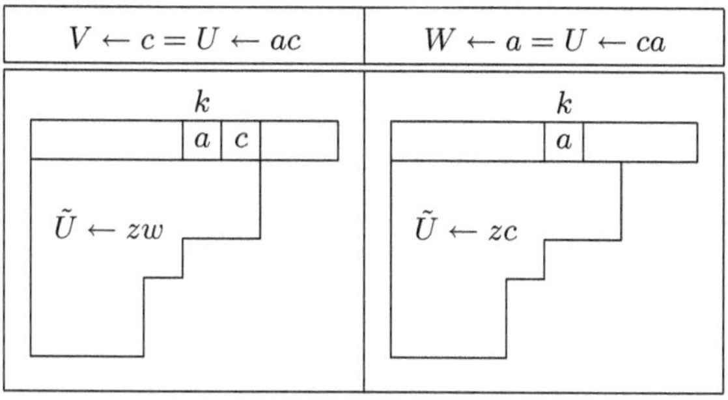

Table C.9. Inserting a, c and c, a into U when $a \le b < c$ and $k = p$.

From $v_{1,k} = a \le b < c = v_{1,k+1}$ it follows that b bumps the letter c (in column $k+1$) of $V \leftarrow c$ into row (2).

From $w_{1,k} = a \le b < c < u_{1,k} \le u_{1,k+1} = w_{1,k+1}$ it follows that b bumps the letter $w = u_{1,k+1}$ into row (2) of $W \leftarrow a$.

The resulting tableaux $V \leftarrow cb = U \leftarrow acb$ and $W \leftarrow ab = U \leftarrow cab$ are shown in table C.10.

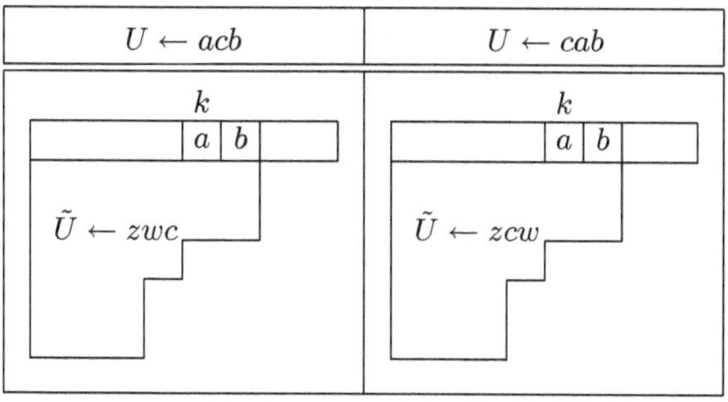

Table C.10. Inserting a, c, b and c, a, b into U when $a \le b < c$ and $k = p$.

It is clear that $c < z \leq w$, because $c < u_{1,p} = z \leq u_{1,p+1} = u_{1,k+1} = w$. This argument is also valid when $z = \infty$, for then $w = \infty$ as well. Therefore part (1) of Proposition (C.4b) implies that $\tilde{U} \leftarrow zwc = \tilde{U} \leftarrow zcw$, and so $U \leftarrow acb = U \leftarrow cab$.

This concludes the proof of Proposition (C.4b), hence of Knuth's theorem (C.3a).

C.5 Littelmann operators on tableaux

Suppose we have $\lambda \in \Lambda^+(n,r)$, $c \in \{1, 2, \ldots, n-1\}$ and a λ-tableau P. The operator $\tilde{f}_c$ does not act on P, but it does act on the word KP. Theorem (C.5b) below will show that there is a unique tableau $\tilde{P}$ such that $K\tilde{P} = \tilde{f}_c(KP)$. It is reasonable to define $\tilde{f}_c(P)$ to be $\tilde{P}$.

In (C.3g), we regarded the entries in the words $KU \mid x_1$ and $K(U \leftarrow x_1)$ as indexed by the r-element set $[\mu] \cup \{(r)\}$. More generally, we can take any ordered r-element set $\mathsf{T} = \{\tau_1, \tau_2, \ldots, \tau_r\}$ such that $\tau_1 < \tau_2 < \cdots < \tau_r$, and use T to index the entries in a word i of length r. This means that if i is a word of length r, we write $i = i_{\tau_1} i_{\tau_2} \ldots i_{\tau_r}$. In this section, it will be convenient to take $\mathsf{T} = [\lambda]$, because $[\lambda]$ indexes the entries of the word KP (see §C.2).

For the moment, let $\mathsf{T} = \{\tau_1, \tau_2, \ldots, \tau_r\}$ be an arbitrary r-element set with $\tau_1 < \tau_2 < \cdots < \tau_r$. Then all the definitions for $\tilde{f}_c$ given in §A.3 translate into definitions for words indexed by T in a trivial manner (we leave it to the reader to make the analogous translations for $\tilde{e}_c$). In case $\mathsf{T} = [\lambda]$ these definitions appear as follows.

First define $\omega_{c,c+1} = \omega : \underline{n} \to \mathbb{Z}$ as in §A.3, so that $\omega(\nu) = 1$, -1 or 0 according as $\nu = c$, $c+1$ or $\nu \notin \{c, c+1\}$.

The map $h_c^{KP} = h^{KP} : [\lambda] \cup \{0\} \to \mathbb{Z}$ is given so that $h^{KP}(0) = 0$, while for any $t \in [\lambda]$ we define

$$\textbf{(C.5a)} \quad h^{KP}(t) := \sum_{(a,b) \leq t} \omega(p_{a,b}),$$

the order $\leq$ being that given by (C.2a).

Let $M = M_c^{KP}$ be the largest element of the set $\{0\} \cup \{\, h_c^{KP} : t \in [\lambda] \,\}$. If $M = 0$ define $\tilde{f}_c(KP) := \infty$, or say that "$\tilde{f}_c(KP)$ is undefined". If $M \neq 0$, let $q = q_c^{KP}$ be the least element t of $[\lambda]$ such that $h^{KP}(t) = M$. Then there must hold $p_{a,b} = c$, where $q = (a,b)$; see (A.3c). In this case we define $\tilde{f}_c(KP)$ to be the word obtained from KP by changing the entry $p_{a,b} = c$ to $c+1$; all other entries in KP are left unchanged.

The next theorem shows that if $\tilde{f}_c(KP) \neq \infty$, it is possible to define a *tableau* $\tilde{f}_c(P)$ in such a way that $K(\tilde{f}_c P) = \tilde{f}_c(KP)$ [3]

---

[3]Some authors identify the tableau P with the word KP, and view theorem (C.5b) as justification of this practice. But in this Appendix we will be cautious (perhaps over-cautious!) and we do not make this identification.

(C.5b) Theorem. *Let $\lambda \in \Lambda^+(n,r)$, $c \in \{1,2,\dots,n-1\}$ and P be a λ-tableau. Using the definitions above, assume $M \neq 0$, and define $q = (a,b)$ to be the least place (in the order (C.2a)) such that $h((a,b)) = M$. We know from (A.3c) that $p_{a,b} = c$. Then we have also*

(1) *If $(a,b+1) \in [\lambda]$, then $p_{a,b+1} \geq c+1$.*

(2) *If $(a+1,b) \in [\lambda]$, then $p_{a+1,b} > c+1$.*

(3) *If we change P to $\tilde{P}$ by changing the entry $p_{a,b} = c$ to $\tilde{p}_{a,b} = c+1$, and leaving unchanged all the other entries in P, we get a λ-tableau $\tilde{P}$ which is standard.*

(4) $K\tilde{P} = \tilde{f}_c(KP)$.

Proof. (1) Since P is standard, $p_{a,b+1} \geq p_{a,b} = c$. If $p_{a,b+1} < c+1$, we would have $p_{a,b+1} = c$. This gives $h_c^{KP}((a,b+1)) = h_c^{KP}((a,b)) + \omega(c) = M+1$, contradicting the definition of M. So there must hold $p_{a,b+1} \geq c+1$.

(2) Since P is standard, we must have $p_{a+1,b} > c$. Unless $p_{a+1,b} > c+1$, we have $p_{a+1,b} = c+1$. We shall show that this leads to a contradiction. Table C.11 shows the rows (a) and $(a+1)$ of P, and their entries in certain columns. Let b'

	$\cdots$	$b'-1$	b'	$\cdots$	b	$b+1$	$\cdots$
a	$\cdots$	$< c$	c	$\cdots$	c	$\geq c$	$\cdots$
$a+1$	$\cdots$	$p_{a+1,b'+1}$	$c+1$	$\cdots$	$c+1$	$> c+1$	$\cdots$

Table C.11. Rows (a) and $(a+1)$ of P.

denote the leftmost of all columns such that $p_{a,b'} = c$. Since $a \geq c+1$, entries in row $(a+1)$ to the right of column (b) are all $> c+1$. The entries in the same row, in columns $(b'),\dots,(b)$, are all equal to $c+1$. This is because such an entry $p_{a+1,b''}$ is left of $p_{a+1,b} = c+1$, and is also $> p_{a,b''} = c$.

From the definition (C.5a) we deduce

(C.5c) $h^{KP}(a,b) = h^{KP}(a+1,b'-1) + X + Y + Y^* + Z,$

where

$$X = \sum_{b' \leq x \leq b} \omega(p_{a+1,x}), \qquad Y = \sum_{b+1 \leq x \leq \lambda_{a+1}} \omega(p_{a+1,x}),$$
$$Y^* = \sum_{1 \leq x \leq b'} \omega(p_{a,x}) \qquad Z = \sum_{b' \leq x \leq b} \omega(p_{a,x}).$$

But for $b+1 \leq x \leq \lambda_{a+1}$ all the entries $p_{a+1,x}$ are $> c+1$, hence all the summands $\omega(p_{a+1,x}) = 0$, therefore $Y = 0$. Similarly $Y^* = 0$ because all the elements $p_{a,x}$ (for $1 \leq x \leq b'-1$) are $< c$. Finally $X + Z = 0$ because $X + Z$ is a sum of pairs $\omega(c) + \omega(c+1) = 0$. Therefore (C.5c) implies that $h^{KP}(a,b) = h^{KP}(a+1,b'-1)$. But this contradicts our definition of (a, b) as the *least* place in $[\lambda]$ such that $h^{KP}(a, b) = M$. This proves part (2) of Theorem (C.5b).

Part (3) is now proved, since (1) and (2) show that $\tilde{P}$ is standard. Then (4) follows.

(C.5d) Example. Let $\lambda = (2, 2, 0, \ldots, 0)$, regarded as an element of $\Lambda^+(n, 4)$ for some $n \geq 4$. Consider the λ-tableau $P = \begin{array}{|c|c|}\hline 2 & 2 \\ \hline 3 & 4 \\ \hline\end{array}$. Then $KP = 3422$. Now

r	1	2	3	4
t	$(2,1)$	$(2,2)$	$(1,1)$	$(1,2)$
$(KP)_r$	3	4	2	2
$h^{KP}(r)$	-1	-1	0	1
$(\tilde{f}_c(KP))_r$	3	4	2	3

Table C.12. Illustration of Theorem (C.5b).

let $c = 2$. Calculate $\tilde{f}_c(KP)$ using table C.12. Notice that we have shown two set $\underline{4}$ and $[\lambda]$, either of which can be used to index the letters of the word KP. We see that $q^{KP} = 4$, or equivalently, $q^{KP} = (1, 2)$. Therefore $\tilde{f}_c(P) = \begin{array}{|c|c|}\hline 2 & 3 \\ \hline 3 & 4 \\ \hline\end{array}$.

C.6 The proof of Proposition B

In this section we shall prove the fact, fundamental for our work, *that the operation KP commutes with all the Littelmann operators $\tilde{f}_c$*. In other words, we shall prove the

Proposition B. *Let $i \in I(n, r)$ and $c \in \{1, 2, \ldots, n-1\}$ such that $\tilde{f}_c(i) \neq \infty$. Then $\tilde{f}_c(KP(i)) \neq \infty$ and*

(C.6a) $\tilde{f}_c(KP(i)) = KP(\tilde{f}_c(i))$.

For $i, j \in I(n, r)$ we write $iK'j$ (respectively, $iK''j$) if i and j are connected by a basic move of type K' (respectively, K''); see (C.3c), (C.3d). The proof of Proposition B is based on the following two lemmas.

(C.6b) Lemma. *Let $i, j \in I(n, r)$, and suppose that j is obtained from i by a basic move, say,*

$$i = (\ldots, i_k, i_{k+1}, \ldots), \quad j = (\ldots, i_{k+1}, i_k, \ldots), \quad i_k < i_{k+1}\,.$$

Then $M^i = M^j$, and there are the following alternatives for q^i and q^j:

(a) *If* $q^i \notin \{k, k+1\}$, *then* $q^j = q^i$.
(b) *If* $q^i = k+1$, *then* $q^j = k$.
(c) *If* $q^i = k$, *then either* $i_{k+1} = c+1$, $iK''j$ *and* $q^j = k+2$, *or* $i_{k+1} \neq c+1$ *and* $q^j = k+1$.

Proof. Set $x := i_k$ and $z := i_{k+1}$, so that $x < z$. We observe that

(i) $h_c^j(\nu) = h_c^i(\nu)$ for all $\nu \neq k$.

This follows directly from the definition (A.3a) of h_c^i and the fact that the words i and j are identical at all places except $\nu = k$ and $\nu = k+1$.

Next we show that

(ii) $M^i = M^j$.

Suppose first that $q^i \neq k$. Then $M^j \geq h_c^j(q^i) = h_c^i(q^i) = M^i$, by (i). Assume that $M^j > M^i$. Then $q^j = k$, by (i). This implies $z = c$ and $x < c$, by (A.3c)(i). It follows that $M^i \geq h_c^i(k+1) = h_c^j(k+1) = h_c^j(k) = M^j$, a contradiction. This shows $M^i = M^j$ in case $q^i \neq k$.

Suppose now $q^i = k$. Then $x = c$, by (A.3c)(i). Hence $M^i \geq M^j$, by (i). Assume that $M^i > M^j$. Then $M^i > h_c^j(k+1) = h_c^i(k+1) = M^i + \omega_c(z)$, which implies $z = c+1$. If $iK'j$, that is, $x < i_{k-1} \leq z$, then $i_{k-1} = c+1$ and thus $h_c^i(k) = h_c^i(k-2) + \omega_c(i_{k-1}) + \omega_c(x) = h_c^i(k-2)$. This contradicts the minimal choice of q^i. If $iK''j$, that is, if $x \leq i_{k+2} < z$, then $i_{k+2} = c$ and thus $M^j \geq h_c^j(k+2) = h_c^i(k+2) = h_c^i(k) = M^i$, again a contradiction. This shows $M^i = M^j$ also in case $q^i = k$, and (ii) is proved.

Now, for the proof of (a), suppose that $q^i \notin \{k, k+1\}$. Assume $q^j \neq q^i$, then $q^j = k$, by (i), and thus $z = c$. It follows that $x < c$ and therefore, by (ii), $h_c^i(k+1) = h_c^j(k) = M^j = M^i$. This implies $q^i < k$, since $q^i \neq k, k+1$, hence also $q^j < k$, by (i)—a contradiction. Part (a) is proved.

Now let $q^i = k+1$. Then $z = c$ and $x < c$, and we get from (ii) that $h_c^j(k) = h_c^j(k+1) = h_c^i(k+1) = M^i = M^j$. It follows that $q^j \leq k$. In fact, by (i), $q^j = k$. This implies (b).

Consider finally the case where $q^i = k$. Then $x = c$, and $q^j \geq k$, by (ii).

Suppose additionally that $z \neq c+1$. Then $z > c+1$, and it follows that $h_c^j(k) < h_c^j(k+1) = h_c^i(k+1) = h_c^i(k) = M^i = M^j$. But this implies $q^j = k+1$.

To conclude, let $z = c+1$. Assume $iK'j$, so that $x < i_{k-1} \leq z$. Then $i_{k-1} = c+1$ and $h_c^i(k) = h_c^i(k-2) + \omega_c(i_{k-1}) + \omega_c(x) = h_c^i(k-2)$. This contradicts the minimal choice of q^i. It follows that $iK''j$ as asserted, that is $x \leq i_{k+2} < z$. Hence $i_{k+2} = c$. Direct verification gives $h_c^j(k) = h_c^i(k) - 2 = M^i - 2 = M^j - 2$, $h_c^j(k+1) = M^j - 1$ and $h_c^j(k+2) = M^j$. Therefore $q^j = k+2$ as claimed in (c).

(C.6c) Lemma. *Let* $i, j \in I(n,r)$, *and suppose* j *is obtained from* i *by a basic move. If* $\tilde{f}_c(i) \neq \infty$, *then* $\tilde{f}_c(j) \neq \infty$, *and* $\tilde{f}_c(j)$ *is obtained from* $\tilde{f}_c(i)$ *by a basic move.*

There is a corresponding statement (and proof), with $\tilde{e}_c$ *replacing* $\tilde{f}_c$.

Proof. Thanks to symmetry in i and j, we may assume that either there exist components $y = i_{k-1}$, $x = i_k$, $z = i_{k+1}$ of i such that

$$(K') \quad i = (\ldots, y, x, z, \ldots), \quad j = (\ldots, y, z, x, \ldots), \quad x < y \le z,$$

or components $x = i_k$, $z = i_{k+1}$, $y = i_{k+2}$ of i such that

$$(K'') \quad i = (\ldots, x, z, y, \ldots), \quad j = (\ldots, z, x, y, \ldots), \quad x \le y < z.$$

Suppose $\tilde{f}_c(i) \ne \infty$. Then $M^j = M^i > 0$, by Lemma (C.6b). This implies that $\tilde{f}_c(j) \ne \infty$ as well.

We now consider the three cases listed in Lemma (C.6b).

Case (a). $q^i \notin \{k, k+1\}$. Then $q^i = q^j$, by Lemma (C.6b). Hence x and z remain unchanged when we apply $\tilde{f}_c$ to i and j. The claim follows directly if y is not changed, either.

Suppose that $y = c$, $iK'j$ and $q^i = k-1$. Then we get[4]

$$\tilde{f}_c(i) = (\ldots, \underline{c+1}, x, z, \ldots), \quad \tilde{f}_c(j) = (\ldots, \underline{c+1}, z, x, \ldots), \quad x < c \le z.$$

However, we have $z \ge c+1$, since in case $z = c$ we get $h^j_c(k) = h^j_c(k-1)+1$, and this contradicts the maximality of $h^j_c(q^j)$. Hence $\tilde{f}_c(i)K'\tilde{f}_c(j)$.

Suppose now that $y = c$, $iK''j$ and $q^i = k+2$. Then we get

$$\tilde{f}_c(i) = (\ldots, x, z, \underline{c+1}, \ldots), \quad \tilde{f}_c(j) = (\ldots, z, x, \underline{c+1}, \ldots), \quad x \le c < z.$$

However, we have $z > c+1$, since in case $z = c+1$ we get $h^i_c(k) = h^i_c(k+2)$, and this contradicts the minimal choice of q^i. Hence $\tilde{f}_c(i)K''\tilde{f}_c(j)$.

Case (b). $q^i = k+1$. Then $q^j = k$ and $z = c$, hence

$$\tilde{f}_c(i) = (\ldots, x, \underline{c+1}, \ldots), \quad \tilde{f}_c(j) = (\ldots, \underline{c+1}, x, \ldots).$$

If $iK''j$, then $x \le y < z < c+1$, therefore $\tilde{f}_c(i)K''\tilde{f}_c(j)$. In case $iK'j$, we get $x < y \le z < c+1$, hence $\tilde{f}_c(i)K'\tilde{f}_c(j)$.

Case (c). $q^i = k$. Here $x = c$, and we need to consider the alternative given in Lemma (C.6b)(c).

Suppose first that $z = c+1$, that $iK''j$ and $q^j = k+2$. Then $y = c$ since $c = x \le y < z = c+1$. Hence

$$\tilde{f}_c(i) = (\ldots, \underline{c+1}, c+1, c, \ldots), \quad \tilde{f}_c(j) = (\ldots, c+1, c, \underline{c+1}, \ldots).$$

We get that $\tilde{f}_c(j)K'\tilde{f}_c(i)$.

The case where $z \ne c+1$ and $q^j = k+1$ remains. Here $z > c+1$ and

$$\tilde{f}_c(i) = (\ldots, \underline{c+1}, z, \ldots), \quad \tilde{f}_c(j) = (\ldots, z, \underline{c+1}, \ldots).$$

Suppose $iK''j$, then $y \ge c+1$ since otherwise $h^i_c(k+2) = h^i_c(k)+1$. Therefore $c+1 \le y < z$ and $\tilde{f}_c(i)K''\tilde{f}_c(j)$. Similarly, if $iK'j$, then $y > c+1$, since otherwise $h^i_c(k-2) = h^i_c(k)$. Hence $c+1 < y \le z$ and $\tilde{f}_c(i)K'\tilde{f}_c(j)$.

---

[4] Those values which were changed by $\tilde{f}_c$ are underlined.

We are now in a position to give the

Proof of Proposition B. From Proposition (C.2c), we get $P(KP(i)) = P(i)$. Hence, by Theorem (C.3a), there exist words

$$i^{(0)},\ i^{(1)},\ \cdots,\ i^{(k-1)},\ i^{(k)} \in I(n,r)$$

such that $i^{(0)} = i$, $i^{(k)} = KP(i)$, and $i^{(\nu)}$ is obtained from $i^{(\nu-1)}$ by a basic move. From Lemma (C.6c), it follows that $\tilde{f}_c(i^{(\nu)}) \neq \infty$ and that $\tilde{f}_c(i^{(\nu)})$ is obtained from $\tilde{f}_c(i^{(\nu-1)})$ by a basic move, for all $\nu \in \{1, \ldots, k\}$.

Applying Theorem (C.3a) again, we get

$$P\Big(\tilde{f}_c(i)\Big) = P\Big(\tilde{f}_c(i^{(0)})\Big) = P\Big(\tilde{f}_c(i^{(k)})\Big) = P\Big(\tilde{f}_c(KP(i))\Big),$$

hence

$(*)$ $KP(\tilde{f}_c(i)) = KP(\tilde{f}_c(KP(i)))$.

There is a standard tableau $\tilde{P}$ such that $K\tilde{P} = \tilde{f}_c(KP(i))$, by Theorem (C.5b)(4). And by Proposition (C.2c)(i), $KP(K\tilde{P}) = K\tilde{P}$. Therefore $(*)$ becomes

$$KP(\tilde{f}_c(i)) = K\tilde{P} = \tilde{f}_c(KP(i))).$$

D

Theorem A and some of its consequences

In what follows, n, r are fixed positive integers.

D.1 Ingredients for the proof of Theorem A

We shall prove Theorem A in the next section, but we must first study some words in $I(n,r)$ which play a special role for the action of the Littelmann operators. To describe these words, we need the following lemma, which is an immediate consequence of the definitions in §A.3.

(D.1a) Lemma. *If $i \in I(n,r)$ and $c \in \{1,\ldots,n-1\}$, then*
(i) $\tilde{f}_c(i) = \infty$ *if and only if*

$$\#\{\nu \leq t : i_\nu = c\} \leq \#\{\nu \leq t : i_\nu = c+1\}$$

for all $t \in \{1,\ldots,r\}$, and
(ii) $\tilde{e}_c(i) = \infty$ *if and only if*

$$\#\{\nu \geq s : i_\nu = c\} \geq \#\{\nu \geq s : i_\nu = c+1\}$$

for all $s \in \{1,\ldots,r\}$.

We are interested in the words which satisfy (i) for all c. So we set

$$\Upsilon := \{\, i \in I(n,r) : \tilde{f}_c(i) = \infty \text{ for all } c \in \{1,\ldots,n-1\}\,\}.$$

Define an operator $W : I(n,r) \to I(n,r)$ by

$$W(i_1 i_2 \ldots i_r) = (n+1-i_1, n+1-i_2, \ldots, n+1-i_r).$$

Then a word i belongs to Υ if and only if $W(i)$ is a "lattice permutation"[1].

[1]This term is rather confusing, because we shall use it for words which may not be permutations! A word $j \in I(n,r)$ is called a *permutation* if $n = r$ and the entries in j are $1, 2, \ldots, r$ in some order. Lattice permutations in this sense are used by D.E. Littlewood in the character theory of the symmetric group $\mathsf{Sym}(r)$ (see [36, page 67]). Lattice permutations in the present sense appear in [40] and [37].

Definition. A *lattice permutation*, in our language, is a word $j \in I(n,r)$ such that

(D.1b) $\#\{\nu \le s : j_\nu = 1\}$
$$\ge \#\{\nu \le s : j_\nu = 2\}$$
$$\ge \cdots$$
$$\ge \#\{\nu \le s : j_\nu = n-1\}$$
$$\ge \#\{\nu \le s : j_\nu = n\},$$

for all $s \in \{1, \ldots, r\}$.

For example, the word $j = 11122132$, an element of $I(3,8)$, is a lattice permutation. The word $i = 33322312$ belongs to Υ, because $W(i) = j$.

Similarly, we are interested in the words which satisfy condition (ii) in Lemma (D.1a) for all c, and we set

$$\mathsf{T} := \left\{ i \in I(n,r) : \tilde{e}_c(i) = \infty \text{ for all } c \in \{1, \ldots, n-1\} \right\}.$$

In (A.3g)(2), the operator $B : I(n,r) \to I(n,r)$ was defined by

$$B(i_1 i_2 \ldots i_{r-1} i_r) = i_r i_{r-1} \cdots i_2 i_1.$$

Thus a word i belongs to T if and only if $B(i)$ is a lattice permutation.

Define an operator $C : I(n,r) \to I(n,r)$ by $C = BW = WB$. Explicitly,

$$C(i_1 i_2 \ldots i_{r-1} i_r) = (n+1-i_r, n+1-i_{r-1}, \ldots, n+1-i_2, n+1-i_1).$$

Remarks.

(i) All these operators have square equal to the identity in $I(n,r)$.

(ii) If $i \in I(n,r)$ and $\mathsf{Sch}(i) = (\lambda(i), P(i), Q(i))$, then $\lambda(i)$ is the shape of i (see §C.1). The operator C preserves shape (i.e. $\lambda(Ci) = \lambda(i)$, see (D.3g)), but the operators B and W do not. For example, using the tables in §E.1, we see that $i = 221$ has shape $(2,1,0)$, but $B(i) = 122$ and $W(i) = 223$ both have shape $(3,0,0)$. However, $C(i) = 322$ has the same shape as i.

(D.1c) Lemma. *The operator C induces a bijection $\mathsf{T} \to \Upsilon$. Hence $|\mathsf{T}| = |\Upsilon|$.*

Proof. Let $i \in \mathsf{T}$. Then $B(i)$ is a lattice permutation, and $W(C(i)) = B(i)$. This shows that $C(i) \in \Upsilon$. Prove similarly that $i \in \mathsf{T}$ implies that $C(i) \in \Upsilon$.

From now on in this section, we fix $\lambda \in \Lambda^+(n,r)$.

The tableaux $\mathbf{T}_\lambda$ and $\mathbf{Z}_\lambda$. Define two λ-tableaux as follows:

(D.1d) $T_\lambda = (T_{s,t})_{(s,t)\in[\lambda]}$ where $T_{s,t} = s$ for all $(s,t) \in [\lambda]$; we denote the word KT_λ by i^λ.

(D.1e) $Z_\lambda = (Z_{s,t})_{(s,t)\in[\lambda]}$ where $Z_{s,t} = n - \beta_t + s$ for all $(s,t) \in [\lambda]$, and β_t denotes the length of column t of Z_λ; we denote the word KZ_λ by i_λ.

Example. If $\lambda = (5,3,2,0,0) \in \Lambda^+(5,10)$, then

$$\textbf{(D.1f)}\quad T_\lambda = \begin{array}{ccccc} 1&1&1&1&1\\ 2&2&2&&\\ 3&3&&& \end{array} \quad\text{and}\quad Z_\lambda = \begin{array}{ccccc} 3&3&4&5&5\\ 4&4&5&&\\ 5&5&&& \end{array}.$$

Notice that T_λ is our old friend from (4.3b), where it is called T_l. It is useful to think of Z_λ as the tableau obtained from T_λ by subjecting it to two successive operations: first reverse each column of T_λ, and secondly replace each entry x in the tableau by $n+1-x$. In our example,

$$T_\lambda = \begin{array}{ccccc} 1&1&1&1&1\\ 2&2&2&&\\ 3&3&&& \end{array} \longrightarrow \begin{array}{ccccc} 3&3&2&1&1\\ 2&2&1&&\\ 1&1&&& \end{array} \longrightarrow \begin{array}{ccccc} 3&3&4&5&5\\ 4&4&5&&\\ 5&5&&& \end{array} = Z_\lambda.$$

Notation. Define $\mathcal{Q}(\lambda)$ to be the set of all (standard) λ-tableaux whose entries are $1, 2, \ldots, r$ in some order.

Recall from (C.1e) that $I(Q, \approx)$ is the set of all words $i \in I(n,r)$ such that $Q(i) = Q$, for each $Q \in \mathcal{Q}(\lambda)$. These sets are the equivalence classes for $\approx$.

(D.1g) Theorem. *Let $Q \in \mathcal{Q}(\lambda)$. Then:*

(i) *There is a unique word $i \in I(n,r)$ such that $Q(i) = Q$ and i belongs to T. Moreover $P(i) = T_\lambda$.*

(ii) *There is a unique word $i \in I(n,r)$ such that $Q(i) = Q$ and i belongs to Υ. Moreover $P(i) = Z_\lambda$.*

(D.1h) Notation. For each $Q \in \mathcal{Q}(\lambda)$, denote the word i in (i) by i^Q, and the word i in (ii) by i_Q.

Proof of Theorem (D.1g). (i) By Schensted's Theorem (B.6a) there is a unique word i such that $P(i) = T_\lambda$ and $Q(i) = Q$. We claim that this i belongs to T.

Let $c \in \{1, 2, \ldots, n\}$. From §A.3, we know that $\tilde{e}_c(i) = \infty$ if and only if the function h_c^i attains its maximum M_c^i at the last place in the word i, i.e. $h_c^i(r) = M_c^i$. Now $h_c^i(r)$ is the sum of the $\omega(i_\nu)$ for $\nu = 1, 2, \ldots, r$ (see (A.3a)). But the r entries in the word $KP(i)$ form a permutation of the r entries of i. Hence $h_c^{KP(i)}(r) = h_c^i(r) = M_c^i$. By Lemma (C.6b) and Proposition (C.3p), the words i and $KP(i)$ give the same maximum, i.e. $M_c^i = M_c^{KP(i)}$. We can calculate $M_c^{KP(i)} = M_c^{KT_\lambda}$ easily; it is $\lambda_c - \lambda_{c+1}$, and it is attained at the last place $(1, \lambda_1)$ of $KP(i)$. Therefore the maximum M_c^i of h_c^i is also attained at the last place of i. This shows that $\tilde{e}_c(i) = \infty$ for all c, hence $i \in \mathsf{T}$.

Now we must prove *uniqueness*: if $j \in I(Q, \approx) \cap \mathsf{T}$, then $j = i$. It is enough to prove that $P(j) = T_\lambda$. We know $\tilde{e}_c(KP(j)) = KP(\tilde{e}_c(j))$, by Proposition B, hence $\tilde{e}_c(KP(j)) = \infty$ for all c. So the height function $h_c^{KP(j)}$ takes its maximum value at "place r", i.e. at place $(1, \lambda_1) \in [\lambda]$.

Consider the last entry t in the first row of $P(j)$; we must show that $t = 1$. If this is false, then $t > 1$. Consider the height functions for $c = t - 1$. The entry in $P(j)$ at place $(1, \lambda_1)$ (which corresponds to place r in the word $KP(j)$) is $c + 1$. So we have $h_c^{KP(j)}(r) < h_c^{KP(j)}(r-1)$, a contradiction.

Hence all entries of the first row of $P(j)$ are equal to 1. Next consider the last entry, t say, in the sth row of $P(j)$. We have $t \geq s$ since $P(j)$ is standard. Suppose $t > s$, and set $c = t - 1$. As before, the height function $h_c^{KP(j)}$ does not take its maximum value at r: it is constant on the letters of rows 1 up to $s - 1$, and its value at the last place of row s, say x, is less than its value at the place immediately preceding this last place. This is a contradiction. We have proved that $P(j) = T_\lambda$, and since we have assumed $Q(j) = Q$, the words j and i must be equal.

If we let λ vary over all partitions in $\Lambda^+(n, r)$, then this shows that $|\mathsf{T}|$ is equal to the number of standard tableaux having entries $1, 2, \ldots, r$ in some order; this is also the total number of $\approx$-classes in $I(n, r)$.

(ii) By Schensted's theorem (B.6a) there is a unique word $i \in I(n, r)$ with $Q(i) = Q$ and $P(i) = Z_\lambda$. Using Lemma (C.6b) and Proposition B, as in the proof of (i), it is quite easy to see that $i \in \Upsilon$.

Therefore each $\approx$-class of words of shape λ contains at least one element of Υ. But as was noted above, if we let λ vary, the number of $\approx$-classes is $|\mathsf{T}|$, which is equal to $|\Upsilon|$ by Lemma (D.1c). This implies that each $\approx$-class contains a unique element of Υ; this must be the word i having $P(i) = Z_\lambda$ and $Q(i) = Q$.

This completes the proof of Theorem (D.1g).

(D.1i) Proposition. $i^\lambda = i^{\mathsf{Q}(\lambda)}$ *and* $i_\lambda = i_{\mathsf{Q}(\lambda)}$.

Proof. Taking $Y = T_\lambda$ in propositions (C.2c) and (C.2h), we get $P(i^\lambda) = T_\lambda$ and $Q(i^\lambda) = \mathsf{Q}^{(\lambda)}$. However (D.1g) and (D.1h) say that $P(i^{\mathsf{Q}(\lambda)}) = T_\lambda$ and $Q(i^{\mathsf{Q}(\lambda)}) = \mathsf{Q}^{(\lambda)}$. Therefore $i^\lambda = i^{\mathsf{Q}(\lambda)}$, by Schensted's theorem (B.6a). A similar proof, using Z_λ in place of T_λ, gives $i_\lambda = i_{\mathsf{Q}(\lambda)}$.

D.2 Proof of Theorem A

We shall now prove the Theorem A described in the introduction (see (A.4a)):

(D.2a) Theorem A. *Let $i, j \in I(n, r)$. Then $i \approx j$ if and only if there is a finite sequence of words (elements of $I(n, r)$):*

$$i(1),\ i(2),\ \ldots,\ i(s)$$

such that $i(1) = i$, $i(s) = j$ and for each adjacent pair $i(\nu)$, $i(\nu + 1)$ ***either*** *there exists an element $c \in \{1, \ldots, n-1\}$ such that $\tilde{f}_c(i(\nu)) = i(\nu + 1)$,* ***or*** *there exists an element $c \in \{1, \ldots, n-1\}$ such that $\tilde{e}_c(i(\nu)) = i(\nu + 1)$.*

It is clear that the "if" part of this theorem is equivalent to the following

(D.2b) Proposition. *If $i, j \in I(n,r)$ and $c \in \{1,\dots,n-1\}$ such that either $\tilde{f}_c(i) = j$, or $\tilde{e}_c(i) = j$, then $Q(i) = Q(j)$.*

Proof. Suppose there is an element $c \in \{1,\dots,n-1\}$ such that $j = \tilde{f}_c(i)$. This implies that $\tilde{f}_c(i) \neq \infty$.

(a) We claim that $Q(i)$ and $Q(j)$ have the same shape. Equivalently, we claim that $P(i)$ and $P(j)$ have the same shape. Let $P(i)$ have shape λ. By Proposition B (see (C.6b)) we know that $KP(j) = KP(\tilde{f}_c(i)) = \tilde{f}_c(KP(i))$. Now take $P = P(i)$ in Theorem (C.5b). This says that there is a λ-tableau $\tilde{P}$ such that $K\tilde{P} = \tilde{f}_c(KP)$. Therefore $KP(j) = K\tilde{P}$. This shows that $P(j)$ has the same shape λ as $\tilde{P}$, which is the shape of $P(i)$. This proves claim (a).

(b) We shall use induction on r to prove that $Q(i) = Q(j)$. If $r = 1$ then i and j are one-letter words, and $Q(i) = Q(j)$ follows from (B.2b). Assume now that $r > 1$. Write $i = i'i_r$ and $j = j'j_r$, where $i' = i_1 \cdots i_{r-1}$ and $j' = j_1 \cdots j_{r-1}$ lie in $I(n, r-1)$.

There is a place $q \in \{1,2,\dots,r\}$ such that $i_q = c$, $j_q = c+1$ and $i_\nu = j_\nu$ for all $\nu \neq q$ (see (A.3e)). We either have $q < r$, or $q = r$. If $q < r$, then $j' = \tilde{f}_c(i')$, hence $Q(i') = Q(j')$ by the induction hypothesis. If $q = r$, then $j' = i'$ and clearly $Q(i') = Q(j')$.

It follows that $Q(i')$ and $Q(j')$ have the same shape, μ say, in either case. Let λ be the shape of $Q(i)$. By (a), λ is also the shape of $Q(j)$. To get $Q(i)$ from $Q(i')$, one puts r into the unique place which, when added to $[\mu]$, gives $[\lambda]$. To get $Q(j)$ from $Q(j') = Q(i')$, one puts r in the unique place which, when added to $[\mu]$, gives $[\lambda]$. Hence $Q(j) = Q(i)$. This completes the proof of Proposition (D.2b); and this proves the "if" part of Theorem A.

Proof of the "only if" part of Theorem A. We assume that $i, j \in I(n,r)$ are such that $Q(i) = Q(j)$; we must prove that there exists a sequence of words $i = i(1), (2), \dots, i(s) = j$ with the properties listed in (D.2a). First make the

(D.2c) Definition. Let $i \in I(n,r)$. We define the *size* of i, denoted $\mathsf{sz}(i)$, by $\mathsf{sz}(i) := i_1 + i_2 + \cdots + i_r$. This is a positive integer, and from the definitions of $\tilde{f}_c$, $\tilde{e}_c$ it is clear that if $\tilde{f}_c(i) \neq \infty$ then $\mathsf{sz}(\tilde{f}_c(i)) = \mathsf{sz}(i)+1$, and if $\tilde{e}_c(i) \neq \infty$ then $\mathsf{sz}(\tilde{e}_c(i)) = \mathsf{sz}(i) - 1$.

We make the convention $\mathsf{sz}(\infty) = 0$.

Let $\lambda \in \Lambda^+(n,r)$ and $Q \in \mathcal{Q}(\lambda)$. Let $w \in I(Q, \approx)$ (see (C.1e)), and let $\mathcal{S}(w)$ denote the set of all words of the form $\tilde{e}_{c_1}\tilde{e}_{c_2} \cdots \tilde{e}_{c_t}(w)$, where $c_1, c_2, \dots, c_t$ are arbitrary elements of $\{1,2,\dots,n-1\}$. We allow that t may be zero, in which case $\tilde{e}_{c_1}\tilde{e}_{c_2} \cdots \tilde{e}_{c_t}(w) = w$. In general, $\tilde{e}_{c_1}\tilde{e}_{c_2} \cdots \tilde{e}_{c_t}(w)$ has size $\mathsf{sz}(w) - t$. Let S be minimal amongst the sizes of the elements of $\mathcal{S}(w)$, and choose an element $w' := \tilde{e}_{c_1}\tilde{e}_{c_2} \cdots \tilde{e}_{c_t}(w)$ of $\mathcal{S}(w)$ of size S (there may be many such w'). Then $\tilde{e}_c(w') = \infty$ for all $c \in \{1,2,\dots,n-1\}$, because if $\tilde{e}_c(w') \neq \infty$, then $\tilde{e}_c(w')$ would be an element of $\mathcal{S}(w)$ of size $S - 1$.

By Proposition (D.2b), all the elements of $\mathcal{S}(w)$ lie in $I(Q,\approx)$. But then Theorem (D.1g) tells us that $w' = \tilde{e}_{c_1}\tilde{e}_{c_2}\cdots\tilde{e}_{c_t}(w) = i^Q$. This implies that $w = \tilde{f}_{c_t}\cdots\tilde{f}_{c_2}\tilde{f}_{c_1}(i^Q)$. In other words, any element $w \in I(Q,\approx)$ can be joined to i^Q by a finite sequence of steps of the form $i(\nu) \xrightarrow{\tilde{f}} i(\nu+1)$; equivalently i^Q can be joined to w by a sequence of steps of the form $i(\nu+1) \xrightarrow{\tilde{e}} i(\nu)$. So given $i,j \in I(Q,\approx)$, we can join i to j by a sequence of the type described in the statement of Theorem A, by first joining i to i^Q and then joining i^Q to j. This completes the proof of Theorem A.

The arguments just given, together with Theorem (D.1g), provide valuable information on the $\approx$-classes. We summarize this in the

(D.2d) Proposition. *Let $\lambda \in \Lambda^+(n,r)$ and $Q \in \mathcal{Q}(\lambda)$. Then:*

(i) *There is a unique word i^Q in $I(Q,\approx)$ lying in T, i.e. such that $\tilde{e}_c(i) = \infty$ for all $c \in \{1,\ldots,n-1\}$. This word is specified by $\mathsf{Sch}(i^Q) = (\lambda, T_\lambda, Q)$.*

(ii) *There is a unique word i_Q in $I(Q,\approx)$ lying in Υ, i.e. such that $\tilde{f}_c(i) = \infty$ for all $c \in \{1,\ldots,n-1\}$. This word is specified by $\mathsf{Sch}(i_Q) = (\lambda, Z_\lambda, Q)$.*

(iii) *The following three conditions on a word $i \in I(n,r)$ are equivalent (i.e. each condition implies the other two):*

(1) $i \in I(Q,\approx)$.

(2) *There exist $c_1,\ldots,c_t \in \{1,\ldots,n-1\}$ such that $i = \tilde{f}_{c_1}\cdots\tilde{f}_{c_t}(i^Q)$.*

(3) *There exist $d_1,\ldots,d_s \in \{1,\ldots,n-1\}$ such that $i = \tilde{e}_{d_1}\cdots\tilde{e}_{d_s}(i_Q)$.*

In (2) and (3), we allow t and s to be $= 0$, respectively. In these cases we interpret $\tilde{f}_{c_1}\cdots\tilde{f}_{c_t}(i^Q)$ to be i^Q and $\tilde{e}_{d_1}\cdots\tilde{e}_{d_s}(i_Q)$ to be i_Q, respectively.

Proof. All the statements above can be deduced easily from Theorem (D.1g), Theorem (D.2a) (i.e. Theorem A) and the proof of Theorem (D.2a).

Weights. Remember (see (A.3g)(3), or §3.1) that the *weight* $\mathsf{wt}(i)$ of a word i is the n-vector $(w_1,\ldots,w_n)$, where for each $\nu \in \underline{n}$, w_ν is the number of $\rho \in \underline{r}$ such that $i_\rho = \nu$. Classical representation theory of $\mathsf{GL}_n(\mathbb{C})$, which can be regarded as a sequel to classical invariant theory, uses weights extensively—they describe the (polynomial) representations K_λ of the diagonal subgroup $T_n(\mathbb{C})$, see §3.2; then these are "induced" to give irreducible (polynomial) representations of $\mathsf{GL}_n(\mathbb{C})$, see the end of Chapter 4.

(D.2e) Remark. It is clear that $\mathsf{wt}(i) = \mathsf{wt}(j)$, if $i,j \in I(n,r)$ are such that $j = i\pi$ for some π in the symmetric group $\mathsf{Sym}(r)$ (the symmetric group is denoted $G(r)$ in §2.1). In particular, $\mathsf{wt}(i) = \mathsf{wt}(KP(i))$ for any $i \in I(n,r)$, because the entries in $KP(i)$ are the same as the entries in i, apart from a place permutation $\pi \in \mathsf{Sym}(r)$.

In the classical representation theory of $\mathsf{GL}_n(\mathbb{C})$, which is essentially the representation theory of the Schur algebra $S(n,r)$, the (isomorphism types of) simple modules are indexed by dominant weights, i.e. by the elements of $\Lambda^+(n,r)$. We shall see in §D.4 that this holds also for the (isomorphism types of) simple modules for the Littelmann algebra $L = L(n,r)$, although the argument is different from that which applies to $S(n,r)$.

The weights of the elements of $I(Q,\approx)$ have properties given in the next proposition.

(D.2f) Proposition. *Let* $\lambda \in \Lambda^+(n,r)$ *and* $Q \in \mathcal{Q}(\lambda)$*, then*

(i) $\mathsf{wt}(i^Q) = \lambda = (\lambda_1, \dots, \lambda_n)$,

(ii) $\mathsf{wt}(i_Q) = (\lambda_n, \dots, \lambda_1)$,

(iii) i^Q *(respectively,* i_Q*) is the only word in* $I(Q,\approx)$ *having weight* $(\lambda_1, \dots, \lambda_n)$ *(respectively,* $(\lambda_n, \dots, \lambda_1)$*), and*

(iv) *the weight* ω *of any word in* $I(Q,\approx)$ *satisfies the inequalities*

$$(\lambda_1, \dots, \lambda_n) \trianglerighteq \omega \trianglerighteq (\lambda_n, \dots, \lambda_1).$$

(If $\xi, \eta \in \Lambda(n,r)$, we write $\xi \trianglerighteq \eta$ to mean that the difference $\xi - \eta$ lies in the set $U = \sum_{\alpha \in \Sigma} \mathbb{Z}_+ \alpha$; see [33, page 3]).

Proof. (i) From (D.1g)(i) we know that $P(i^Q) = T_\lambda$. Therefore $\mathsf{wt}(i^Q)$ is the same as the weight of KT_λ (see Remark (D.2e)). It is very easy to see that $\mathsf{wt}(KT_\lambda) = (\lambda_1, \dots, \lambda_n)$.

(ii) In the same way, we deduce from (D.1g)(ii) that $\mathsf{wt}(i_Q)$ is the same as the weight $(u_1, \dots, u_n)$ of KZ_λ. So for each $\delta \in \underline{n}$, u_δ is the number of pairs $(s,t) \in [\lambda]$ such that $n - \beta_t + s = \delta$. For each $t \in \underline{n}$, there is exactly one entry δ in column t of Z_λ, if and only if $1 \le \beta_t - (n - \delta)$. Therefore u_δ equals the number of columns of Z_λ of lengths greater than or equal to $n + 1 - \delta$. But this number is $\lambda_{n+1-\delta}$.

(iii) and (iv) Let i be any word in $I(Q,\approx)$. By (D.2d) we know that there exist integers $c_1, \dots, c_t$ in $\{1, \dots, n-1\}$ such that $i = \tilde{f}_{c_1} \cdots \tilde{f}_{c_t}(i^Q)$. From (A.3g)(3), we know that $\mathsf{wt}(i) = \mathsf{wt}(i^Q) - \alpha_{c_1,c_1+1} - \cdots - \alpha_{c_t,c_t+1}$. Therefore $i \trianglelefteq i^Q$; moreover the case $\mathsf{wt}(i) = \mathsf{wt}(i^Q) = (\lambda_1, \dots, \lambda_n)$ occurs only if $t = 0$ i.e. only if $i = i^Q$. A similar argument shows that the weight of any word $i \in I(Q,\approx)$ is $\trianglerighteq \mathsf{wt}(i_Q)$, with equality only if $i = i_Q$.

D.3 Properties of the operator C

First we want to understand how the action of C is related to the action of the Littelmann operators.

Comparing the height functions of i and Ci. Fix $i \in I(n,r)$, and consider the height function h_c^i for some $c \in \{1, 2, \dots, n-1\}$. This depends on the i_ν which are equal to c or $c+1$. The operator C turns c and $c+1$ into $n-c+1$ and $n-c$, respectively. This suggests comparing h_c^i with h_{n-c}^{Ci}.

Take some $s \in \{1, \dots, r\}$, then by definition

(D.3a) $h_c^i(s) = \#\{\nu \le s : i_\nu = c\} - \#\{\nu \le s : i_\nu = c+1\}.$

Now write $Ci = j_1 \dots j_r$ for a moment and consider

(D.3b) $h^{Ci}_{n-c}(r-s) = \#\{\rho \le r-s : j_\rho = n-c\}$
$$-\#\{\rho \le r-s : j_\rho = n-c+1\}.$$

We have $j_\rho = n - i_{r-\rho+1} + 1$ for all $\rho \in \{1, \dots, r\}$. Furthermore, $n - i_\nu + 1 = n - c$ if and only if $i_\nu = c+1$, and $n - i_\nu = n - c$ if and only if $i_\nu = c$, for all $\nu \in \{1, \dots, r\}$. So (D.3b) gives

(D.3c) $h^{Ci}_{n-c}(r-s) = \#\{\nu \ge s+1 : i_\nu = c+1\} - \#\{\nu \ge s+1 : i_\nu = c\}.$

By using the notation:

$$\Pi_b := \#\{\nu \in \Pi : i_\nu = b\},$$

for every subset Π of $\{1, \dots, s\}$, and every element b of $\underline{n}$, formula (D.3c) becomes

(i) $h^{Ci}_{n-c}(r-s) = -\{s+1, \dots, r\}_c + \{s+1, \dots, r\}_{c+1}$.

Also, by definition of the height function h^i_c, we have

(ii) $h^i_c(s) = \{1, \dots, s\}_c - \{1, \dots, s\}_{c+1}$.

If we subtract (i) from (ii) we get

(D.3d) *If* $i \in I(n,r)$ *and* $Y = h^i_c(r)$, *then* $h^i_c(s) - h^{Ci}_{n-c}(r-s) = Y$ *for all* $s \in \{0, \dots, r\}$.

Example. Let $n = 3$, $r = 5$, $c = 1$, and consider $i = 22111 \in I(n,r)$. Then $Ci = 33322$ and $n - c = 2$. The height functions h^i_1 and h^{Ci}_2 are

$$\Big(\ h^i_1(0),\ h^i_1(1),\ h^i_1(2),\ h^i_1(3),\ h^i_1(4),\ h^i_1(5)\Big) = (\ \ 0, -1, -2, -1,\ \ 0,\ \ 1),$$

$$\Big(h^{Ci}_2(5), h^{Ci}_2(4), h^{Ci}_2(3), h^{Ci}_2(2), h^{Ci}_2(1), h^{Ci}_2(0)\Big) = (-1, -2, -3, -2, -1,\ \ 0).$$

Note that h^i_1 has the maximum at place r and h^{Ci}_2 has maximum value zero.

(D.3e) Lemma. *Let* $c \in \{1, 2, \dots, n-1\}$. *Then, for each* $i \in I(n,r)$, *we have* $C(\tilde{e}_c(i)) = \tilde{f}_{n-c}(Ci)$ *and* $C(\tilde{f}_c(i)) = \tilde{e}_{n-c}(Ci)$.

Proof. Since C^2 is the identity, the second part follows from the first. We prove the first part.

By (D.3d), we have a geometric description of how the height functions are related: given $h = h^i_c$, then to find $\tilde{h} = h^{Ci}_{n-c}$, one reflects the graph of h in the vertical line $x = r$, and translates it in the "y-axis" direction so that $\tilde{h}(0) = 0$. Explicitly, let $s + t = r$ and $Y = h^i_c(r)$, then

$$h^{Ci}_{n-c}(t) = h^i_c(s) - Y.$$

Since $\tilde{h}$ is a reflection about a vertical line, the last maximum of h_c^i (at place $\bar{q}$), becomes the first maximum of h_{n-c}^{Ci} (at place $r-\bar{q}$). Furthermore, if $\bar{q} = r$ then the maximum of $\tilde{h}$ is zero. This shows that if $\tilde{e}_c(i) = \infty$ then $\tilde{f}_{n-c}(Ci) = \infty$.

Assume now that $\bar{q} < r$. By (A.3c) we know that $i_{\bar{q}+1} = c+1$, and $\tilde{e}_c(i)$ is obtained from i by replacing $i_{\bar{q}+1} = c+1$ by c. We get the word

$$C(\tilde{e}_c(i)) = \cdots (n-c+1)(n-i_{\bar{q}}+1)\cdots$$

where the letters shown are at places $r-\bar{q}$ and $r-\bar{q}+1$.

Now consider $C(i) = \cdots (n-c)(n-i_{\bar{q}}+1)\cdots$ where the letters shown are at places $r-\bar{q}$ and $r-\bar{q}+1$. We know that $\tilde{h}$ assumes its maximum at place $r-\bar{q}$ for the first time. So $\tilde{f}_{n-c}(Ci)$ replaces the letter $n-c$ at place $r-\bar{q}$ by $n-c+1$. Hence $\tilde{f}_{n-c}(Ci)$ is equal to $C(\tilde{e}_c(i))$.

From the definition of C we see immediately the following.

(D.3f) Lemma. *If a word $i \in I(n,r)$ has weight $\mu = (\mu_1,\ldots,\mu_n)$, then the word Ci has weight $(\mu_n,\ldots,\mu_1)$.*

We want to show now:

(D.3g) Lemma. *The operator C preserves the shape.*

Proof. Let $i \in I(n,r)$ and $Q(i) = Q$, and suppose Q has shape λ.

Assume first that $i = i^Q$, then $C(i) = i_R$ for some standard tableau R. By (D.2f) we can identify the shapes of the words i^Q and i_R from their weights. The weight of i^Q is λ, hence the weight of $C(i^Q)$ is $(n^{\lambda_1},(n-1)^{\lambda_2},\ldots)$. So i_R also has shape λ.

In general, by (D.2d)(iii), there are $c_1,\ldots,c_t \in \{1,2,\ldots,n-1\}$ such that

$$i = \tilde{f}_{c_1}\cdots\tilde{f}_{c_t}(i^Q).$$

From (D.3e) it follows that $C(i) = \tilde{e}_{n-c_1}\cdots\tilde{e}_{n-c_t}(Ci^Q)$. But we have already seen that Ci^Q has shape λ; now Proposition B implies that $C(i)$ also has shape λ.

(D.3h) Remark. The operator C does not preserve the Q-symbol in general. But it gives a pairing on the set of standard tableaux of the same shape.

D.4 The Littelmann algebra $L(n,r)$

(D.4a) Let V be an n-dimensional vector space over a field F with basis $v_1,\ldots,v_n$. Then the r-fold tensor product $V^{\otimes r}$ has basis $\{v_i : i \in I(n,r)\}$. (In §§1–6, $V^{\otimes r}$ is called $E^{\otimes r}$.) For each $\alpha \in \Sigma$, where $\alpha = \alpha_{c,c+1}$ (see §A.3), we let $\tilde{f}_c$ and $\tilde{e}_c$ act on the tensor space by linear maps, defining

$$\tilde{f}_c v_i := v_{\tilde{f}_c i}, \qquad \tilde{e}_c v_i := v_{\tilde{e}_c i}$$

and using linear extension. We set $v_\infty := 0$.

Let $L = L(n,r)$ be the subalgebra of $\mathsf{End}_F(V^{\otimes r})$ generated by these linear maps $\tilde{f}_c$ and $\tilde{e}_c$, for $c \in \{1, 2, \ldots, n-1\}$. This algebra will be called the *Littelmann algebra.*

(D.4b) As an F-space, the Littelmann algebra L is spanned the set of all monomials $m = m_1 m_2 \ldots m_t$ of lengths $t \geq 1$, where each m_τ is either $\tilde{f}_c$ or $\tilde{e}_c$ (for some $c \in \{1, 2, \ldots, n-1\}$).

We do *not* include the monomial $m = 1_{\mathsf{End}_F V^{\otimes r}}$ of length zero. But it may happen that L does contain this element (see Proposition (D.4e), below).

(D.4c) An element $H \in \mathsf{End}_F(V^{\otimes r})$ will often be described by its matrix $(H_{i,j})_{i,j \in I(n,r)}$, whose entries $H_{i,j} \in F$ are defined by the equations

(D.4d) $Hv_j = \sum_{i \in I(n,r)} H_{i,j} v_i$, all $j \in I(n,r)$.

We often identify H with its matrix $(H_{i,j})_{i,j \in I(n,r)}$, and we often identify $\tilde{f}_c$, $\tilde{e}_c$ with the elements of $\mathsf{End}_F(V^{\otimes r})$ defined by (D.4a).

(D.4e) Proposition. *L has an identity element, viz. D_S, the diagonal matrix having $(D_S)_{i,i} = 1, 0$ according as $i \in S$ or not; here $S := I(n,r) \setminus (\Upsilon \cap \mathsf{T})$.*

Reminder: from §D.1 we have

$$\Upsilon = \{\, i \in I(n,r) \, : \, \tilde{f}_c(i) = \infty \text{ for all } c \in \{1, 2, \ldots, n-1\} \,\}$$

and

$$\mathsf{T} = \{\, i \in I(n,r) \, : \, \tilde{e}_c(i) = \infty \text{ for all } c \in \{1, 2, \ldots, n-1\} \,\}.$$

Proof of Proposition (D.4e). For any subset A of $I(n,r)$, define D_A to be the element of $\mathsf{End}_F(V^{\otimes r})$ whose matrix with respect to the basis $\{v_i : i \in I\}$ is diagonal, and $(D_A)_{ii} = 1$ or 0, according as $i \in A$ or $i \notin A$. The following facts are easily checked.

(i) $D_{I(n,r)}$ is the identity element of $\mathsf{End}_F(V^{\otimes r})$.

(ii) For any c, the matrix of $\tilde{f}_c \tilde{e}_c$ is equal to $D_{Z(c)}$, where $Z(c)$ is the set of all i such that $\tilde{e}_c(i) \neq \infty$. Similarly $\tilde{e}_c \tilde{f}_c = D_{Y(c)}$, where $Y(c)$ is the set of all i such that $\tilde{f}_c(i) \neq \infty$.

(iii) If $A, B \subseteq I(n,r)$, then $D_A D_B = D_{A \cap B}$ and $D_{A \cup B} = D_A + D_B - D_{A \cap B}$.

(iv) If $A_1, \ldots, A_w \subseteq I(n,r)$ such that $D_{A_t} \in L$ for all $t = 1, 2, \ldots, w$, then $D_A \in L$ where $A = A_1 \cup A_2 \cup \ldots \cup A_w$.

Now check that $I(n,r) \setminus \mathsf{T} = \bigcup_c Z(c)$ and $I(n,r) \setminus \Upsilon = \bigcup_c Y(c)$, hence

$$S = I(n,r) \setminus (\Upsilon \cap \mathsf{T}) = \bigcup_c Z(c) \cup \bigcup_c Y(c).$$

The partition $I(n,r) = S \cup (\Upsilon \cap \mathsf{T})$ of $I(n,r)$ allows us to decompose each linear operator $H \in \mathsf{End}_F(V^{\otimes r})$ in matrix form as

$$H = \begin{pmatrix} H^{(1,1)} & H^{(1,2)} \\ H^{(2,1)} & H^{(2,2)} \end{pmatrix},$$

with $H^{(1,1)} \in \mathsf{End}_F(S,S)$, $H^{(1,2)} \in \mathsf{Hom}_F(\Upsilon \cap \mathsf{T}, S)$, $H^{(2,1)} \in \mathsf{Hom}_F(S, \Upsilon \cap \mathsf{T})$ and $H^{(2,2)} \in \mathsf{End}_F(\Upsilon \cap \mathsf{T})$. For each $c \in \{1, 2, \ldots, n-1\}$, it is easy to verify the following facts.

(v) If $H = \tilde{e}_c$, or if $H = \tilde{f}_c$, then $H^{(1,2)}$, $H^{(2,1)}$, $H^{(2,2)}$ are all zero matrices; also $\tilde{f}_c$ is the transpose of $\tilde{e}_c$.

(vi) If $H \in L$, then $H^{(1,2)}$, $H^{(2,1)}$, $H^{(2,2)}$ are all zero and

$$H = \begin{pmatrix} H^{(1,1)} & 0 \\ 0 & 0 \end{pmatrix}.$$

(vii) D_S is the matrix shown, with $H^{(1,1)}$ the identity matrix.

Proposition (D.4e) follows from these facts.

(D.4f) Example. If $n = r = 2$, then $I(n,r) = \{11, 12, 21, 22\} = S \cup (\Upsilon \cap \mathsf{T})$, where $S = \{11, 12, 22\}$, and $\Upsilon \cap \mathsf{T} = \{21\}$. We have

$$\tilde{e}_1 = \begin{pmatrix} 0 & 1 & 0 & 0 \\ 0 & 0 & 1 & 0 \\ 0 & 0 & 0 & 0 \\ 0 & 0 & 0 & 0 \end{pmatrix}, \quad \tilde{f}_1 = \begin{pmatrix} 0 & 0 & 0 & 0 \\ 1 & 0 & 0 & 0 \\ 0 & 1 & 0 & 0 \\ 0 & 0 & 0 & 0 \end{pmatrix}, \quad D_S = \begin{pmatrix} 1 & 0 & 0 & 0 \\ 0 & 1 & 0 & 0 \\ 0 & 0 & 1 & 0 \\ 0 & 0 & 0 & 0 \end{pmatrix}.$$

D.5 The modules M_Q

Let $\lambda \in \Lambda^+(n,r)$. For each standard tableau Q in $\mathcal{Q}(\lambda)$ we define M_Q to be the subspace of $V^{\otimes r}$ which has F-basis all v_i such that $Q = Q(i)$, that is, all i in $I_\lambda(Q, \approx)$. By Proposition B, this is an L-submodule of $V^{\otimes r}$.

We get therefore a direct sum decomposition of the tensor space $V^{\otimes r}$ into L-submodules

$$V^{\otimes r} = \bigoplus_{\lambda \in \Lambda^+(n,r)} \bigoplus_{Q \in \mathcal{Q}(\lambda)} M_Q.$$

(D.5a) $M_Q = L v_{i^Q} = L v_{i_Q}$.

This follows from (D.2d).

For $z = \sum_i \xi_i v_i \in V^{\otimes r}$, define the *support* of z to be

$$\mathsf{supp}(z) = \{\, i \in I(n,r) \, : \, \xi_i \neq 0 \,\}.$$

(D.5b) Lemma. *If* $z = \sum_i \xi_i v_i$ *and* $c \in \{1, 2, \ldots, n-1\}$ *such that* $\tilde{e}_c z = 0$, *then* $\mathsf{supp}(z)$ *lies in the set* $I(n,r) \setminus Z(c)$.

Proof. Let $U' := \{\, i : \tilde{e}_c(i) \neq \infty \,\} = Z(c)$ and $U'' := \{\, i : \tilde{e}_c(i) = \infty \,\}$. Then $z = z' + z''$, where $z' = \sum_{i \in U'} \xi_i v_i$ and $z'' = \sum_{i \in U''} \xi_i v_i$.

We have by assumption

$$(*) \ \sum_{i \in U'} \xi_i \, v_{\tilde{e}_c(i)} = \tilde{e}_c z = 0.$$

But for each $i \in U'$, $\tilde{e}_c(i) \neq \infty$, hence $\tilde{f}_c \tilde{e}_c(i) = i$. Applying $\tilde{f}_c$ to $(*)$, we get $0 = \sum_{i \in U'} \xi_i v_i$. But the v_i are linearly independent, so all the ξ_i (for $i \in U'$) are zero. Therefore $z' = 0$ which shows that $z = z''$ has support in $U'' = I(n,r) \setminus Z(c)$.

(D.5c) Corollary. *If* $z \in V^{\otimes r}$ *is annihilated by all* $\tilde{e}_c$, $c \in \{1, 2, \ldots, n-1\}$, *then* $\mathsf{supp}(z)$ *lies in* $\bigcap_c I(n,r) \setminus Z(c) = I(n,r) \setminus \bigcup_c Z(c) = \mathsf{T}$.

Similarly, if $z \in V^{\otimes r}$ is annihilated by all $\tilde{f}_c$, $c \in \{1, 2, \ldots, n-1\}$, then $\mathsf{supp}(z)$ lies in $I(n,r) \setminus \bigcup_c Y(c) = \Upsilon$. There may be a word $i \in I(n,r)$ such that v_i is annihilated by the algebra L. According to Proposition (D.2d), we must have $i^Q = i = i_Q$ in this case, and the $\approx$-class of i consists of i alone. From (D.2f), the shape λ of Q has the property $(\lambda_1, \ldots, \lambda_n) = (\lambda_n, \ldots, \lambda_1)$, hence $\lambda = (k^n)$ (and $r = nk$). In this case $M_Q = Fv_i$. There may be more than one such $i \in I_\lambda(n,r)$. For example, $I_{(2,2)}(2,4)$ contains $i = (2211)$ and $j = (2121)$ in $\Upsilon \cap \mathsf{T}$.

As a consequence, we have to allow L-modules which are not unital. An L-module M is then defined to be an F-space on which L acts by linear transformations so that $x(ym) = (xy)m$, for all $x, y \in L$ and $m \in M$. The L-module M is defined to be simple (= irreducible) if *either*

(1) $LM = 0$ and M is a simple F-space, that is, has F-dimension 1, *or*
(2) $M \neq 0$, the element D_S of L acts as the identity on M, and M has no L-submodules except M and $\{0\}$.

We aim to show that the modules M_Q are simple as L-modules. To do so, it is helpful to exploit a subalgebra of L.

(D.5d) The involutory anti-automorphism J of $\mathbf{L} = \mathbf{L}(\mathbf{n}, \mathbf{r})$. At this point we find a strong similarity with the involutory anti-automorphism J of $S(n,r)$ defined in §2.7.

Define a symmetric bilinear map $\langle \ , \ \rangle$ on $V^{\otimes r}$ by the rule $\langle v_i, v_j \rangle = \delta_{i,j}$ for all $i, j \in I(n,r)$. Given $H \in \mathsf{End}_F(V^{\otimes r})$, we defined in (D.4c), (D.4d) its matrix $(H_{i,j})$. Now define $J(H) \in \mathsf{End}_F(V^{\otimes r})$ by the rule: the matrix of $J(H)$ is the transpose of the matrix of H.

By (D.4d), $H_{i,j} = \langle Hv_j, v_i \rangle$ for all $i, j \in I(n,r)$. Replacing H by $J(H)$, we get $J(H)_{i,j} = \langle J(H)v_j, v_i \rangle$. But by definition, $J(H)_{i,j} = H_{j,i} = \langle Hv_i, v_j \rangle$. Therefore $\langle Hv_i, v_j \rangle = \langle J(H)v_j, v_i \rangle = \langle v_i, J(H)v_j \rangle$ for all $i, j \in I(n,r)$. Equivalently,

(D.5e) $\langle Hv, w \rangle = \langle v, J(H)w \rangle$ for all $v, w \in V^{\otimes r}$.

We may use (D.5e) as a definition of $J(H)$ when H is given.

From the elementary properties of transposed matrices we see that the linear map $J : \mathsf{End}_F V^{\otimes r} \to \mathsf{End}_F V^{\otimes r}$ is involutory (i.e. J^2 is the identity) and is an anti-automorphism (i.e. $J(H_1H_2) = J(H_2)J(H_1)$ for all H_1, H_2). But from our present viewpoint the important fact is

(D.5f) For all $c \in \{1, 2, \ldots, n-1\}$ there holds $J(\tilde{f}_c) = \tilde{e}_c$ and $J(\tilde{e}_c) = \tilde{f}_c$.

These facts follow from the properties stated in (A.3g)(4). We leave the details as an exercise for the reader. But from (D.5f) we see that J maps L into itself, so it gives a map $J : L \to L$ which is an involutory anti-automorphism of the algebra $L = L(n,r)$.

(D.5g) Take a (total) order $\leq$ on the set $I(n,r)$ such that $\mathsf{sz}(i) \leq \mathsf{sz}(j)$ implies that $i \leq j$ for all words i, j in $I(n,r)$. Using such an order, the matrix of $\tilde{e}_c$ is upper triangular, and the matrix of $\tilde{f}_c$ is lower triangular.

We have therefore:

(D.5h) Corollary. *Let L^+ be the subalgebra of L, generated by the elements $\tilde{e}_c$, $c \in \{1, 2, \ldots, n-1\}$. Then L^+ is nilpotent.*

(D.5i) Lemma. *The module M_Q is simple.*

Note that this holds for arbitrary fields F.

Proof. This is clear if M_Q is an 1-dimensional module which is annihilated by L, so we assume that this is not the case.

We fix $i = i^Q$, then v_i generates M_Q as an L-module, by (D.2a).

Let $0 \neq x \in M_Q$, it suffices to show that v_i lies in the L-submodule generated by x.

To do so, we consider the L^+-submodule of M_Q generated by x. By (D.5h), this submodule contains some non-zero element z such that $\tilde{e}_c z = 0$ for all c. By (D.5c), the support of z lies in T. But it also lies in $I(Q, \approx)$ since $z \in M_Q$. It follows that z is a scalar multiple of v_i, by (D.2d)(i). We assumed z is non-zero, hence v_{i^Q} lies in the submodule generated by x.

(D.5j) It follows by a well-known theorem on finite dimensional algebras (or on rings with minimum condition; see e.g. [11, Theorem (25.2), page 164]), that L is semisimple.

Furthermore, every unital simple L-module M occurs as a submodule (and hence as a summand) of $V^{\otimes r}$. Take any non-zero element $x \in M$. Then $M = Lx$, and the map $\theta : L \to M$ which takes $u \mapsto ux$ is an epimorphism of L-modules. But since L is semisimple, it follows that θ maps some simple submodule N of L isomorphically onto M. And the simple submodules of L are submodules of $V^{\otimes r} = \bigoplus_{\lambda \in \Lambda^+(n,r)} \bigoplus_{Q \in \mathcal{Q}(\lambda)} M_Q$ (see the displayed formula, above (D.5a)).

This shows that every unital simple L-module is isomorphic to M_Q for some $Q \in \mathcal{Q}(\lambda)$, for some $\lambda \in \Lambda^+(n,r)$.

To classify the simple L-modules we must find out when M_Q and M_R are isomorphic.

D.6 The λ-rectangle

Fix $\lambda \in \Lambda^+(n,r)$ and use the following notation:

(D.6a) $\mathcal{P}(\lambda) = \{P_1, P_2, \dots, P_{d_\lambda}\}$ is the set of all standard λ-tableaux whose entries all lie in $\underline{n}$, and

(D.6b) $\mathcal{Q}(\lambda) = \{Q_1, Q_2, \dots, Q_{f_\lambda}\}$ is the set of all standard λ-tableaux whose entries are $\{1, 2, \dots, r\}$ in some order (see D.1).

Definition. If $P \in \mathcal{P}(\lambda)$ and $Q \in \mathcal{Q}(\lambda)$, let $P : Q$ denote[2] the word $i \in I(n,r)$ such that $P(i) = P$ and $Q(i) = Q$. In the notation of B.7,

(D.6c) $P : Q = \mathsf{M}(\lambda, P, Q) = \mathsf{Sch}^{-1}(\lambda, P, Q)$.

It is useful to display the set of words of shape λ in the following "λ-rectangle"

(D.6d)
$$\begin{array}{cccc} P_1 : Q_1 & P_1 : Q_2 & \cdots & P_1 : Q_{f_\lambda} \\ P_2 : Q_1 & P_2 : Q_2 & \cdots & P_2 : Q_{f_\lambda} \\ \vdots & \vdots & & \vdots \\ P_{d_\lambda} : Q_1 & P_{d_\lambda} : Q_2 & \cdots & P_{d_\lambda} : Q_{f_\lambda} \end{array}$$

This rectangle has the following properties:

(D.6e) (i) Every element of $I_\lambda(n,r)$ appears once and only once in (D.6d) (see (B.6a)).

(ii) The hth row $\{P_h : Q_1, \dots, P_h : Q_{f_\lambda}\}$ is the $\sim$-class $I_\lambda(P_h, \sim)$, for each $h \in \{1, \dots, d_\lambda\}$ (see (C.1d)).

(iii) The kth column $\{P_1 : Q_k, \dots, P_{d_\lambda} : Q_k\}$ is the $\approx$-class $I_\lambda(Q_k, \approx)$, for each $k \in \{1, \dots, f_\lambda\}$ (see (C.1e)).

---

[2] Not to be confused with the bideterminant $(T_i : T_j)$ defined in (4.3a).

(D.6f) From now on we shall arrange the notation in the λ-rectangle (D.6d) so that $P_1 = T_\lambda$ and $Q_1 = \mathsf{Q}^{(\lambda)}$. Recall from (C.2i) that $\mathsf{Q}^{(\lambda)} \in \mathcal{Q}(\lambda)$ has the property: an element $i \in I_\lambda(n,r)$ has $Q(i) = \mathsf{Q}^{(\lambda)}$ if and only if $i = KP(i)$.

D.7 Canonical maps

Fix $\lambda \in \Lambda^+(n,r)$ again. The entries $P : Q$ in (D.6d) are elements of $I(n,r)$. From now on we shall make the

(D.7a) Convention. When convenient, we shall regard each $i \in I(n,r)$ as the element v_i of $V^{\otimes r}$.

With this convention, the column of the rectangle (D.6d) corresponding to a given $Q \in \mathcal{Q}(\lambda)$ is a basis of the L-module M_Q (see §D.5).

(D.7b) Definition. If $Q, R \in \mathcal{Q}(\lambda)$, then the F-linear map $\gamma_{Q,R} : M_Q \to M_R$ which takes $P_h : Q \mapsto P_h : R$ for each $P_h \in \mathcal{P}(\lambda)$, is the *canonical map* from M_Q to M_R.

Since any two columns in (D.6d) have the same length, the canonical map is an F-linear isomorphism. It is clear that $\gamma_{Q,R}\gamma_{S,Q} = \gamma_{S,R}$ and $\gamma_{Q,R} = (\gamma_{R,Q})^{-1}$, for all $Q, R, S \in \mathcal{Q}(\lambda)$. Our ambition in this section is to prove that any canonical map is an isomorphism of L-modules (see (D.7f) and (D.7i)), and that any L-homomorphism $M_Q \to M_R$ is a scalar multiple of the canonical map $\gamma_{Q,R}$ (see (D.7h)).

(D.7c) Lemma. *Let $Q \in \mathcal{Q}(\lambda)$ and $P \in \mathcal{P}(\lambda)$, and let $i = P : Q$. Then*

$$KP(i) = P : \mathsf{Q}^{(\lambda)} = \gamma_{Q,\mathsf{Q}^{(\lambda)}}(i).$$

In other words, the operation $i \to KP(i)$ is achieved (for $i \in I_\lambda(n,r)$) by the canonical map $\gamma_{Q,\mathsf{Q}^{(\lambda)}}$.

Proof. By (C.3p) one may make a sequence of basic moves joining i to $KP(i)$. Since basic moves do not change P-symbols, we know that $KP(i) = P : Q'$ for some $Q' \in \mathcal{Q}(\lambda)$. But $KP(i)$ equals $KP(KP(i))$, hence its Q-symbol is $\mathsf{Q}^{(\lambda)}$ (see (C.2i)). Therefore $KP(i) = P : \mathsf{Q}^{(\lambda)}$.

Notice that this holds for any i in column Q of (D.6d), and in particular it holds for $\tilde{f}_c(i)$, for any $c \in \{1, \ldots, n-1\}$. By Proposition B (see C.6) we have $\tilde{f}_c(KP(i)) = KP(\tilde{f}_c(i))$, and in our case this gives

(D.7d) $\tilde{f}_c(P : \mathsf{Q}^{(\lambda)}) = KP(\tilde{f}_c(i))$

or, as a commutative diagram,

$$\text{(D.7e)}\qquad \begin{array}{ccc} KP(i) & \xleftarrow{\ \gamma\ } & i \\ \downarrow & & \downarrow \\ KP(\tilde{f}_{c(i)}) & \xleftarrow{\ \gamma\ } & \tilde{f}_{c(i)} \end{array}$$

where $\gamma = \gamma_{Q,\mathsf{Q}^{(\lambda)}}$ and the vertical arrows indicate action of $\tilde{f}_c$. Thus the action of $\tilde{f}_c$ commutes with γ, when applied to any i in the Q-column of (D.6d). In the same way, one has a diagram like (D.7e), with $\tilde{e}_c$ replacing $\tilde{f}_c$. Then we can replace $\tilde{f}_c$ in (D.7e) by any element of L and still have a commutative diagram, since the elements $\tilde{f}_c$ and $\tilde{e}_c$, $c \in \{1, \ldots, n-1\}$ generate L as F-algebra. So we get a

(D.7f) Corollary to (D.7c). *For each pair Q, R of tableaux in $\mathcal{Q}(\lambda)$, the canonical map $\gamma_{Q,R} : M_Q \to M_R$ is an isomorphism of L-modules.*

Proof. The argument above shows that the corollary holds if $R = \mathsf{Q}^{(\lambda)}$, and this gives the general case, since $\gamma_{Q,R} = \gamma^{-1}_{R,\mathsf{Q}^{(\lambda)}} \gamma_{Q,\mathsf{Q}^{(\lambda)}}$.

(D.7g) Lemma. *Suppose Q, R are both tableaux whose entries are $1, 2, \ldots, r$ in some order (possibly of different shapes). If $\psi : M_Q \to M_R$ is a homomorphism of L-modules, then $\psi(v_{i^Q}) = \alpha v_{i^R}$ for some $\alpha \in F$.*

Proof. Let $z = \psi(v_{i^Q})$, then $\tilde{e}_c(z) = \psi(\tilde{e}_c(v_{i^Q})) = 0$ for all $c \in \{1, 2, \ldots, n-1\}$. Therefore $\mathsf{supp}(z) \subseteq \mathsf{T}$, by Corollary (D.5c). But also $\mathsf{supp}(z) \subseteq I(R, \approx)$ since $z \in M_R$, hence $\mathsf{supp}(z) \subseteq \mathsf{T} \cap I(R, \approx) = \{i^R\}$ (see (D.2d)); this means that $z = \alpha v_{i^R}$ for some $\alpha \in F$.

(D.7h) Corollary. *Suppose Q, R are tableaux whose entries are $1, 2, \ldots, r$ in some order. If Q, R have the same shape then $\mathsf{Hom}_L(M_Q, M_R) = F\gamma_{Q,R}$.*

Proof. If $\psi \in \mathrm{Hom}_L(M_Q, M_R)$, i.e. if $\psi : M_Q \to M_R$ is an L-homomorphism, then $\psi(v_{i^Q}) = \alpha v_{i^R}$ for some $\alpha \in F$. But in the present case we know that $\gamma_{Q,R} : M_Q \to M_R$ also is an L-homomorphism, by (D.7f). Therefore we have L-homomorphisms ψ and $\alpha\gamma_{Q,R}$ which take v_{i^Q} to the same element αv_{i^R}. Since v_{i^Q} is an L-generator of M_Q, by (D.5a), the map ψ is equal to $\alpha\gamma_{Q,R}$.

The module M_Q is simple, and $M_Q \cong M_{\mathsf{Q}^{(\lambda)}}$. For each λ, let $M_\lambda = M_{\mathsf{Q}^{(\lambda)}}$.

(D.7i) Lemma. *Let $\lambda, \mu \in \Lambda^+(n, r)$, then $M_\lambda \cong M_\mu$ if and only if $\lambda = \mu$.*

Proof. Suppose that there is an isomorphism $\psi : M_\lambda \to M_\mu$ of L-modules. Then $\psi(v_{i^\lambda}) = \alpha v_{i^\mu}$ for some $\alpha \in F$, by (D.7g). (Note that $M_\lambda = M_{\mathsf{Q}^{(\lambda)}}$ and, by (D.1i), $i^{\mathsf{Q}^{(\lambda)}} = i^\lambda$ and $i^{\mathsf{Q}^{(\mu)}} = i^\mu$.)

If we apply repeatedly $\tilde{f}_1$'s to i^λ then we replace each time the last 1 by a 2, and we can do this $\lambda_1 - \lambda_2$ times, and the next time we get zero. Similarly if we apply $\tilde{f}_1$ to i^μ repeatedly, then we can do this $\mu_1-\mu_2$ times before we get zero. The isomorphism shows now that $\lambda_1-\lambda_2=\mu_1-\mu_2$. The same argument with $\tilde{f}_2$ shows that $\lambda_2-\lambda_3=\mu_2-\mu_3$, and so on. Both λ and μ have degree r which forces $\lambda=\mu$.

Exercise 1. Let $\lambda=(5,4,2)$. Find $\tilde{f}_1 v_{i^\lambda}$ and verify that $\tilde{f}_1^2 v_{i^\lambda}=v_\infty=0$. Find also $\tilde{f}_2^t v_{i^\lambda}$ for $t=1,2,3$.

Exercise 2. If $\lambda=(k,\dots,k)$, where $kn=r$, we have

$$T_\lambda \;=\; \begin{array}{|c|c|c|c|}\hline 1 & 1 & \cdots & 1 \\ \hline 2 & 2 & \cdots & 2 \\ \hline \vdots & \vdots & & \vdots \\ \hline k & k & \cdots & k \\ \hline \end{array}\,.$$

Check by direct calculation that $\tilde{f}_c(KT_\lambda)=\infty=\tilde{e}_c(KT_\lambda)$ for all c.

D.8 The algebra structure of $L(n,r)$

Each M_λ has endomorphism algebra F. It follows now that L is isomorphic to the direct sum of matrix algebras,

$$L\cong\bigoplus_\lambda M_{d_\lambda}(F),$$

where the sum is taken over all $\lambda\in\Lambda^+(n,r)$ with $\lambda\neq(k^n)$, and where d_λ denotes the dimension of M_λ. This follows from the Frobenius–Schur theorem, see [11, Theorem 27.8, page 183], but we shall give a direct proof that the representation $L\to\mathsf{End}_F(M_\lambda)$ afforded by the simple module M_λ is surjective, for all $\lambda\in\Lambda^+(n,r)$, $\lambda\neq(k^n)$.

The problem is to give elements of L which realize the "matrix units". Fix $\lambda\in\Lambda^+(n,r)$. The module M_λ is simple, it has F-basis $\{\,v_i\,:\,i\in I(\lambda)\,\}$ where we set

$$I(\lambda):=I(\mathsf{Q}^{(\lambda)},\approx)=\{\,i\in I_\lambda\,:\,KP(i)=i\,\}$$

(see (C.2i)). An element Φ of $\mathsf{End}_F(M_\lambda)$ is regarded as a matrix $(\Phi_{ij})_{i,j\in I(\lambda)}$ in the usual way,

$$\Phi(v_j)=\sum_{i\in I(\lambda)}\Phi_{ij}v_i,\ \text{all } j\in I(\lambda).$$

Monomials in $\tilde{f}_c,\tilde{e}_c$ are often identified with the matrices of the linear transformations which they determine on M_λ.

(D.8a) Lemma. *Suppose Φ is a monomial in the $\tilde{e}_c$ and $\tilde{f}_c$, $c \in \{1, \dots, n-1\}$. Let $(\Phi_{ij})_{i,j \in I(\lambda)}$ be the matrix of Φ restricted to M_λ.*

(i) *Each row or column of the matrix has at most one non-zero entry (which is then equal to 1).*

(ii) *If $(\Phi_{ij})_{i,j \in I(\lambda)}$ has rank 1, then it is a matrix unit.*

Proof. For each $i \in I(\lambda)$, $\Phi(v_i)$ is either zero, or a basis element. So all but at most one entries of each column are zero, and if there is a non-zero entry then it is equal to 1.

This implies part (i), since if $\Phi = m_1 \cdots m_t$, then $J(\Phi) = J(m_t) \cdots J(m_1)$ is also a monomial. But $(J(\Phi))$ is the transpose of (Φ) (see (D.5d)).

Part (ii) follows.

(D.8b) We want to show that every element of $\mathsf{End}_F(M_\lambda)$ can be represented by some element of L, i.e. that the map $L \to \mathsf{End}_F(M_\lambda)$ is surjective. And we would like to do this by showing that each "matrix unit" $E_{i,j} \in M_{d_\lambda}(F)$ can be represented by some polynomial in the $\tilde{f}_c, \tilde{e}_c$.

It is enough to prove that E_{i_λ, i^λ} can be represented by a monomial in $\mathcal{L}$. Namely, if $s, t \in I(\lambda)$ then by (D.2d) there are monomials p and q in $\tilde{f}$'s and $\tilde{e}$'s with $p(v_{i_\lambda}) = v_s$ and $q(v_t) = v_{i^\lambda}$. The linear map $pE_{i_\lambda, i^\lambda}q$ of $V^{\otimes r}$ has rank at most 1 (the rank of E_{i_λ, i^λ}), so by (D.8a) it is equal to E_{st}, and this is then also represented by an element in L.

The L-module $M_\lambda = M_{\mathsf{Q}^{(\lambda)}}$ has basis $\{ v_i \ : \ i \in I(\lambda) = I(\mathsf{Q}^{(\lambda)}, \approx) \}$. The d_λ elements of $I(\lambda)$ can be arranged (see (D.1g) and (D.6d)) as

$$i(1) = i^{\mathsf{Q}^{(\lambda)}} = i^\lambda, \quad i(2), \quad \dots, \quad i(d_\lambda - 1), \quad i(d_\lambda) = i_{\mathsf{Q}^{(\lambda)}} = i_\lambda.$$

(D.8c) Proposition.

(i) *There are $c(1), \dots, c(b) \in \{1, \dots, n-1\}$ such that $\Phi := \tilde{f}_{c(b)} \cdots \tilde{f}_{c(1)}$ maps i^λ to i_λ.*

(ii) *The number b in (i) is given by $\mathsf{sz}(i^\lambda) + b = \mathsf{sz}(i_\lambda)$.*

(iii) *If $d(1), \dots, d(s) \in \{1, \dots, n-1\}$ are such that $\tilde{f}_{d(s)} \cdots \tilde{f}_{d(1)} i^\lambda \neq \infty$, then $s \leq b$.*

(iv) *$\Phi(v_{i(a)}) = 0$, for all $a \in \{2, 3, \dots, d_\lambda\}$.*

This shows that the matrix of Φ on M_λ is the matrix unit E_{i_λ, i^λ}.

Proof. (i) is a direct application of (D.2d)(iii)(2). We recall the proof, because it brings up useful information.

The idea is to apply operators $\tilde{f}_{c(1)}, \tilde{f}_{c(2)}, \dots$ in succession to the word i^λ, in such a way that, for each $t = 1, 2, \dots$

$$\tilde{f}_{c(t)} \cdots \tilde{f}_{c(1)} i^\lambda \neq \infty.$$

The word $\tilde{f}_{c(t)} \cdots \tilde{f}_{c(1)} i^\lambda$ has size $\mathsf{sz}(i^\lambda) + t$, by (D.2c). The sizes of words in $I(n, r)$ are bounded by rn. Hence, however we choose $c(1), c(2), \dots,$ we must reach b such that $\tilde{f}_{c(b)} \cdots \tilde{f}_{c(1)} i^\lambda \neq \infty$, but $z = \tilde{f}_{c(b)} \cdots \tilde{f}_{c(1)} i^\lambda$ has the property $\tilde{f}_c(z) = \infty$ for all $c \in \{1, \ldots, n-1\}$. This implies $z \in \Upsilon$; however $z \in I(\lambda) = I(\mathsf{Q}^{(\lambda)}, \approx)$, by (D.2b), hence $z = i_\lambda$ by (D.2d)(ii). This proves parts (i) and (ii) of (D.8c).

To prove part (iii), note that the argument above (replace t by s) shows that, applying further operators $\tilde{f}_{d(s+1)}, \tilde{f}_{d(s+2)}, \ldots, \tilde{f}_{d(b')}$ to $\tilde{f}_{d(s)} \cdots \tilde{f}_{d(1)} i^\lambda$ if necessary, we must reach b' such that

$$\tilde{f}_{d(b')} \cdots \tilde{f}_{d(s+1)} \tilde{f}_{d(s)} \cdots \tilde{f}_{d(1)} i^\lambda = i_\lambda.$$

Taking the size of each side of this equation, we get $\mathsf{sz}(i^\lambda) + b' = \mathsf{sz}(i_\lambda)$; this shows that $b' = b$. Therefore $s \leq b' = b$.

(iv) If $a \in \{2, 3, \ldots, d_\lambda\}$, then by (D.2d)(iii), $i(a) = \tilde{f}_{d(1)} \cdots \tilde{f}_{d(u)} i^\lambda$ for some $d(1), \ldots, d(u) \in \{1, \ldots, n-1\}$, and $u \geq 1$. But this implies that $\Phi(i(a)) = \tilde{f}_{c(b)} \cdots \tilde{f}_{c(1)} \tilde{f}_{d(u)} \cdots \tilde{f}_{d(1)} i^\lambda$. If this were $\neq \infty$, it would contradict (iii), since $b + u > b$. Therefore $\Phi(i(a)) = \infty$, hence $\Phi(v_{i(a)}) = 0$.

Remarks.

(i) In (D.8c)(i), there may be several ways of choosing $c(1), \ldots, c(b)$ so that $\Phi = \tilde{f}_{c(b)} \ldots \tilde{f}_{c(1)}$ maps i^λ to i_λ. But by (ii) the length b of any such sequence is always the same, namely $b = \mathsf{sz}(i_\lambda) - \mathsf{sz}(i^\lambda)$.

(ii) For any $\Phi \in L$, the matrix of $J(\Phi)$ is the transpose of the matrix of Φ. This is true by definition if the matrices are defined in terms of the natural basis $\{ v_i : i \in I(n, r) \}$ of $V^{\otimes r}$ (see (D.5d)), hence it is true also for the matrices defined in terms of the basis $\{ v_i : i \in I(\lambda) \}$ of M_λ. Therefore, if $\Phi = \tilde{f}_{c(b)} \ldots \tilde{f}_{c(1)}$ as in (D.8c), then the map $J(\Phi) = \tilde{e}_{c(1)} \ldots \tilde{e}_{c(b)}$ has matrix E_{i^λ, i_λ}.

Example (see chapter E). . Take $\lambda = (2, 1, 0)$ and $\mathsf{Q}^{(\lambda)} = \begin{array}{|c|c|}\hline 1 & 3 \\ \hline 2 & \\ \cline{1-1}\end{array}$. We can take

$$i(1) = 211, \quad i(2) = 311, \quad i(3) = 312, \quad i(4) = 322, \quad i(5) = 323.$$

Another possibility is to take

$$i(1) = 211, \quad i(2) = 212, \quad i(3) = 213, \quad i(4) = 313, \quad i(5) = 323.$$

D.9 The character of M_λ

The basis $\{ v_i : i \in I(Q, \approx) \}$ for M_Q consists of eigenvectors for the diagonal matrices in the general linear group $\mathsf{GL}(n, F)$. Hence M_Q has a formal character (as defined in §3.4), also when the field F is finite.

Explicitly, let $(M_Q)^\alpha = \xi_\alpha M_Q$ (see §3), then the formal character of M_Q is by definition

$$\Phi_{M_Q}(X_1, \ldots, X_n) = \sum_{\alpha \in \Lambda(n,r)} \mathsf{dim}(M_Q)^\alpha \, X_1^{\alpha_1} \cdots X_n^{\alpha_n}.$$

The P-symbol preserves weights, hence $\mathsf{dim}(M_Q)^\alpha = \mathsf{dim}(M_\lambda)^\alpha$, where λ is the shape of Q. Therefore $\Phi_{M_Q} = \Phi_{M_\lambda}$, that is, the formal character of M_Q depends only on the shape of Q.

Let V_λ be the "Weyl module" associated to λ; this is a module for the Schur algebra, see section 5.2. The following is due to P. Littelmann, in far more generality [35, Introduction].

(D.9a) Corollary. *The modules M_λ and V_λ have the same formal character.*

Proof. In (5.4a) we saw that $(V_\lambda)^\alpha$ has F-basis indexed by standard λ-tableaux of weight α. We also know from the characterisation of M_λ given above that $(M_\lambda)^\alpha$ has basis v_i labelled by standard λ-tableaux of weight α. Hence M_λ has the same formal character as V_λ.

Note that it follows that $\Phi_{M_\lambda} = \Phi_{M_\mu}$ if and only if $\lambda = \mu$. (This is also visible directly, by considering the "highest terms" of the formal characters.)

D.10 The Littlewood–Richardson Rule

Suppose λ and μ are partitions with $\lambda \in \Lambda^+(n,r)$, $\mu \in \Lambda^+(n,s)$. Then

$$\Phi_{V_\lambda} \cdot \Phi_{V_\mu} = \sum_\nu c^\nu_{\lambda,\mu} \Phi_{V_\nu}.$$

The coefficients $c^\nu_{\lambda,\mu}$ are non-negative integers, and the Littlewood–Richardson rule is a combinatorial rule for computing these integers. As we have seen, the L-module M_λ has the same formal character as the Schur algebra module V_λ. Then we have

$$\Phi_{M_\lambda} \cdot \Phi_{M_\mu} = \sum_\nu c^\nu_{\lambda,\mu} \Phi_{M_\nu}.$$

Here the sum is taken over all $\nu \in \Lambda^+(n, r+s)$. This leads to the following combinatorial description of the coefficients.

(D.10a) *Let $\mathcal{W}$ be the set of words $i \in I(n, r+s)$ of the form $i = jk$ with the following properties:*

(a) *$KP(j) = j$ and $P(j)$ has shape λ,*

(b) *$k = i^\mu$, and*

(c) *the reverse $B(i)$ of i is a lattice permutation of weight ν.*

Then $c^\nu_{\lambda,\mu}$ is equal to $\#\mathcal{W}$, the number of elements of $\mathcal{W}$.

A number of proofs of (D.10a) exist, and Littelmann gives a wide ranging generalization to cover any complex symmetrizable Kac-Moody Lie algebra [35, Introduction]. The proof we give is the special case which applies to $\mathsf{gl}(n)$ or to GL_n.

Proof. By definition M_λ is a direct summand of $V^{\otimes r}$, and M_μ is a direct summand of $V^{\otimes s}$. Then $M_\lambda \otimes M_\mu$ is a direct summand of $V^{\otimes(r+s)}$, as a vector space, since $v_i \otimes v_j = v_{ij}$. It is invariant under the linear maps $\tilde{f}_c$ and $\tilde{e}_c$. For example, $\tilde{f}_c(v_{ij})$ is either $v_{\tilde{f}_c(i)j}$ or $v_{i\tilde{f}_c(j)}$, or zero; and each of these belong again to $M_\lambda \otimes M_\mu$.

Furthermore, $M_\lambda \otimes M_\mu$ is the direct sum of L-modules M_R for some standard tableaux R with shapes in $\Lambda^+(n, r+s)$.

Therefore $c^\nu_{\lambda,\mu}$ is precisely the number of such R such that M_R occurs as a direct summand in $M_\lambda \otimes M_\mu$ and $M_R \cong M_\nu$. Each M_R contains a unique "highest weight vector" v_i such that $\tilde{e}_c(i) = \infty$ for all c, namely the basis vector for $i = i^R$.

Hence $c^\nu_{\lambda\mu}$ is equal to the number of words $i = jk$ where

(i) i belongs to T (see §D.1), and $P(i)$ has shape ν;
(ii) $k = KP(k)$, and $P(k)$ has shape μ;
(iii) $KP(j) = j$ and $P(j)$ has shape λ.

We know $i \in \mathsf{T}$ if and only if $B(i)$ is a lattice permutation (see (D.1b)). For $i \in \mathsf{T}$, the shape is the same as the weight (see (D.2f)). Furthermore, if $B(i)$ is a lattice permutation then so is $B(k)$, and since $k = KP(k)$, we have $k = i^\mu$. This completes the proof of (D.10a).

(D.10b) Very often, the Littlewood–Richardson rule is stated in a different form. It says that the coefficient $c^\nu_{\lambda,\mu}$ is equal to the size of the set $\mathcal{C}$ of standard (skew) tableaux T of shape $\nu \setminus \mu$ and of weight λ such that the word $w(T)$ is a lattice permutation. Here the word $w(T)$ is obtained by reading T from right to left and from rows $1, 2, 3, \ldots$.

We will now show directly that $\#\mathcal{C} = \#\mathcal{W}$, by means of a bijection from $\mathcal{W}$ onto $\mathcal{C}$.

Suppose i belongs to $\mathcal{W}$, where $i = jk$ as in (D.10a). Then always $k = i^\mu$, and we must consider the tableau $P(j)$. Let r_{ts} be the number of times the letter s occurs in row t of $P(j)$; since $P(j)$ is standard, row t of $P(j)$ starts with some letter $\geq t$, and it has the form

$$t^{r_{tt}}(t+1)^{r_{t,t+1}} \ldots\ n^{r_{tn}},\ t \geq 0$$

Write the multiplicities r_{st} as an upper triangular matrix:

$$U = \begin{pmatrix} r_{11} & r_{12} & & \cdots \\ & r_{22} & r_{23} & \cdots \\ & & r_{33} & \cdots \\ \vdots & & & \end{pmatrix}.$$

By transposing this matrix, we can define a skew tableaux $T = \psi(U)$, depending on $i = jk$, as follows: The tth row of T starts at position $(t, \mu_t + 1)$ and has the multiplicities taken from the tth column of U, that is, row t is

$$t^{r_{tt}}(t-1)^{r_{t-1,t}} \dots 1^{r_{1t}}.$$

The associated word is then

$$w(T) = 1^{r_{11}}(2^{r_{22}}1^{r_{12}})(3^{r_{33}}2^{r_{23}}1^{r_{13}})\cdots$$

We will show that T belongs to $\mathcal{C}$, and that the map $\psi : U \to T$ is a bijection between $\mathcal{W}$ and $\mathcal{C}$.
(1) The word j has weight $\nu \setminus \mu$ if and only if for each s, the sum of the entries in column s of the matrix U is equal to $\nu_s - \mu_s$. This means for the skew tableau T that the sum of the entries in row s is equal to $\nu_s - \mu_s$, for each s, that is T has shape $\nu \setminus \mu$.
(2) The tableau $P(j)$ has shape λ provided row t of $P(j)$ has λ_t entries, for each t, that is

$$\sum_{v \geq t} r_{tv} = \lambda_t\,.$$

This is equivalent with saying that the skew tableau $w(T)$ has weight λ.
(3) The tableau $P(j)$ is standard if and only if

$$\sum_{y=s+1}^{v+1} r_{s+1,y} \leq \sum_{y=s}^{v} r_{s,y}.$$

for all $s \geq 1$ and all $v \geq s$. This means for the word $w(T)$ that in each initial section the number of entries equal to s is $\geq$ the number of entries equal to $s+1$, for each $s \geq 1$. That is, $P(j)$ is standard if and only if $w(T)$ is a lattice permutation.
(4) The word $B(i)$ is of the form

$$(1^{\mu_1}2^{\mu_2}\cdots)(\cdots x^{r_{1x}}\cdots 2^{r_{12}}1^{r_{11}})(\cdots x^{r_{2x}}\cdots 2^{r_{22}})(\cdots x^{r_{3x}}\cdots 3^{r_{33}})\cdots$$

This is a lattice permutation if and only if for each $s \geq 1$ and each v

$$\mu_s + \sum_{y=1}^{v} r_{ys} \geq \mu_{s+1} + \sum_{y=1}^{v+1} r_{y,s+1}.$$

This is equivalent with T being standard.

Combining (1) to (4), we see that if $i = jk \in \mathcal{W}$ and if U is the matrix encoding j, then the skew tableau $T = \psi(U)$ belongs to $\mathcal{C}$. Conversely if we start with some $T \in \mathcal{C}$, then T is the transpose of a matrix U, and this encodes a word $i = jk$ in $\mathcal{W}$. So ψ is a bijection.

D.11 Lascoux, Leclerc and Thibon

This is a brief summary of Chapter 6 of the collective work "Algebraic combinatorics on words" [38]. This chapter is called "The plactic monoid", and its authors are A. Lascoux, B. Leclerc and J.-Y. Thibon. We refer to this chapter, and to its authors, as LLT. Our main purpose is to show that LLT prove facts which imply Theorem A and Proposition B (see (D.11h)).

Reference numbers for sections, propositions, etc. in LLT are enclosed in square brackets (so that, for example, [6.1] stands for [38, 6.1]).

(D.11a) The background of LLT is work of M. P. Schützenberger, which expresses the combinatoric background of work by A. Young, G. de B. Robinson, D. E. Littlewood, etc. on the representation theory of the finite symmetric group.

(D.11b) Words and tableaux. In LLT the set of all words on the alphabet $A = \{1, \ldots, n\}$ is denoted A^*. So in our language, $A^* = \bigcup_{r \geq 0} I(n, r)$.

In LLT (page 3), a tableau[3] is a word i in A^* such that $i = KP$ for some standard tableau P in the sense of section B.1. For example, $i = 544135$ is a tableau, because $i = KP$ for $P = \begin{array}{|c|c|c|}\hline 1 & 3 & 5 \\ \hline 4 & 4 \\ \cline{1-2} 5 \\ \cline{1-1}\end{array}$. If we know that i is a tableau, the corresponding tableau P (which LLT call its *planar representation*) is uniquely defined. The shape λ of i is, by definition, the shape of P. In the example above, the shape is $\lambda = (3, 2, 1, 0, 0)$.

(D.11c) In [6.1] the Schensted algorithm is described. It takes each word i to a tableau $KP(i)$. We can take $P(i)$ to be the tableau defined in (B.4b), (B.4c). The *equivalence* $\sim$ on A^* is defined in [6.2, bottom of page 4]: if i, j are words, then $i \sim j$ means $KP(i) = KP(j)$. The *equivalence* $\equiv$ on A^* is defined on page 5 to be the equivalence on A^* generated by *basic moves* (see (C.3c), (C.3d) and [6.2.3, 6.2.4]. LLT do not use the term "basic move".)

(D.11d) Knuth's theorem (C.3a), [6.2.5] says that $\equiv$ coincides with $\sim$. This is proved in [6.2], elegantly and economically, by a theorem of C. Greene [21]. Greene's theorem itself is also proved in [6.2]. Now the main

(D.11e) Definition (see [6.2.2]). The *plactid monoid* $\mathsf{Pl}(A) := A^*/\!\sim$ is the quotient of A^* by $\sim$. Elements of $\mathsf{Pl}(A)$ are the $\sim$-classes, or "plactic classes" in A^*.

[3] We write tableau (underlined) for a word which is a "tableau in the sense of LLT". A tableau (not underlined) is a standard tableau in the sense of section B.1 of this Appendix. Later in LLT a tableau KP and its planar representation P are often identified.

Knuth shows that $\sim$ is compatible with the product of words: if u, u', v, v' are words, then $u \sim u'$ and $v \sim v'$ implies $uu' \sim vv'$ (see [34, Corollary, page 724]). Product of words is by concatenation, so that $uu' = u \mid u'$; see (A.3g)(6).

Therefore $\mathsf{Pl}(A)$ is a monoid (i.e. a semigroup with identity): the product of the $\sim$-class of u with the $\sim$-class of u', is defined to be the $\sim$-class of uu'. If u is any word, then $u \sim KP(u)$ (see (C.3p), [6.2.3]). Every $\sim$-class contains exactly one tableau; see Theorem [6.2.5].

(D.11f) A main theme in LLT is that it is often useful to "lift" a symmetric polynomial to $\mathbb{Z}[\mathsf{Pl}(A)]$. Suppose that M is any monoid. Then $\mathbb{Z}[M]$, which is the free $\mathbb{Z}$-module with M as $\mathbb{Z}$-basis, is a ring. In case $M = A^*$ we can identify the ring $\mathbb{Z}[A^*]$ with the tensor ring $T(V) = \mathbb{Z} \oplus V \oplus (V \otimes V) \oplus \cdots$ over the free $\mathbb{Z}$-module $V = \mathbb{Z}\nu_1 \oplus \cdots \oplus \mathbb{Z}\nu_n$, by identifying each word $i = i_1 \cdots i_r \in A^*$ with the tensor product $\nu_i = \nu_{i_1} \otimes \cdots \otimes \nu_{i_r}$ (compare with (D.4a)). Yet another interpretation of $\mathbb{Z}[A^*]$ is as the ring of all polynomials (over $\mathbb{Z}$) in non-commuting variables $\nu_1, \ldots, \nu_n$; here one regards every tensor product $\nu_i = \nu_{i_1} \otimes \cdots \otimes \nu_{i_r}$ as the monomial $\nu_{i_1} \cdots \nu_{i_r}$.

Now suppose that $\xi_1, \ldots, \xi_n$ are commuting variables. Then there is an epimorphism of rings $\kappa : \mathbb{Z}[A^*] \to \mathbb{Z}[\xi_1, \ldots, \xi_n]$ which takes $\nu_\sigma \mapsto \xi_\sigma$ for all $\sigma \in \{1, \ldots, n\}$. And this map factors through the map $\pi : \mathbb{Z}[A^*] \to \mathbb{Z}[\mathsf{Pl}(A)]$ induced by the natural epimorphism $A^* \to \mathsf{Pl}(A)$; this means that $i \sim j$ implies $\kappa(i) = \kappa(j)$. (It is enough to check this in case i is connected to j by a basic move.) So there exists a ring epimorphism $\eta : \mathbb{Z}[\mathsf{Pl}(A)] \to \mathbb{Z}[\xi_1, \ldots, \xi_n]$ such that $\kappa = \eta\pi$. In section [6.4] LLT define a "plactic Schur function" S_λ in $\mathbb{Z}[\mathsf{Pl}(A)]$ which is mapped by η onto the classical Schur function in the variables $\xi_1, \ldots, \xi_n$ (see remark (iii) in section 3.5). Then they deduce the Littlewood–Richardson rule from an identity in $\mathbb{Z}[\mathsf{Pl}(A)]$ (see Theorem [6.4.5]).

(D.11g) Returning to section [6.3]; LLT define Schensted's Q-symbol. So for any $i \in A^*$, one defines the tableau $Q(i)$ (or more correctly the tableau $KQ(i)$) which is a byproduct of the sequence of tableaux $P(i_1), P(i_1 i_2), \ldots$ which is used to make $P(i)$; see the example (B.4c), or the example in LLT (page 7). By its construction, $Q(i)$ is what LLT call a "standard" tableau, i.e. if $i \in I(n,r)$, then the entries of $Q(i)$ are the numbers $1, 2, \ldots, r$ in some order. The shape of $Q(i)$ is the shape λ of $P(i)$. LLT prove the Robinson–Schensted theorem [6.3.1], which says that the map $\rho : i \mapsto (P(i), Q(i))$ induces a bijection from the set $I_\lambda(n,r)$ of all words i of given shape λ (see §C.1) to the set $\mathsf{Tab}(\lambda, A) \times \mathsf{STab}(\lambda)$. (In our notation, $\mathsf{Tab}(\lambda, A) = \mathcal{P}(\lambda)$ and $\mathsf{STab}(\lambda) = \mathcal{Q}(\lambda)$; see (D.6a) and (D.6b).) This is essentially the theorem (B.6a) which says the map Sch is bijective. It is proved in the same way, by constructing the inverse map ρ^{-1}.

The rest of section [6.3] is devoted to applications to representations of the symmetric group $S(n)$. A permutation σ of $\{1, \ldots, n\}$ is regarded as a word $\sigma = \sigma_1 \cdots \sigma_n$ of length n. Then G. de B. Robinson discovered and Schützenberger proved the theorem [6.3.3]: $Q(\sigma) = P(\sigma^{-1})$. LLT give a short proof of this fact, and also generalize it to obtain, for any word i, a description of $Q(i)$ as $P(\sigma^{-1})$ for a certain permutation σ constructed from i (see [6.3.7]). Then a further generalization, gives them a generating function for the number d_λ of plactic classes of given weight λ (see [6.3.10]). Notice that d_λ appears in the "λ-rectangle" (D.6d).

(D.11h) In section [6.5], the set of all $i \in A^*$ for which $Q(i)$ is a given "standard" tableau Q is called a *coplactic class.* In our terminology (see section C.1) this is the $\approx$-class $I_\lambda(Q, \approx)$, where $\approx$ is the equivalence relation on A^* defined in (A.4b): $i \approx j$ means $Q(i) = Q(j)$. (LLT do not give a symbol for $\approx$.)

In order to give "structure" to the coplactic classes, LLT introduce three operations on words (which then induce linear operations on $\mathbb{Z}[A^*]$). For a given $c \in \{1, \ldots, n-1\}$, the LLT operators are called e_c, f_c, σ_c. We shall see in (D.11i) that e_c, f_c are just the Littelmann operators $\tilde{e}_c$, $\tilde{f}_c$ defined in section A.3. We do not have the operator σ_c in the Appendix, but it is used extensively in the latter part of LLT.

Proof of Theorem A. Theorem [6.5.1(i)] says that if θ is either e_c or f_c, then $Q(\theta i) = Q(i)$ for any word i such that $\theta i \neq 0$. This is Proposition (D.2b); it is the "if" part of Theorem A. The "only if" part follows from Proposition [6.5.2(i)]; one defines a graph Γ (called the *Littelmann graph* in section E.2) to have for vertices all words $i \in A^*$, with arrow $i \xrightarrow{c} j$ where $f_c i = j$. If i, j are such that $i \approx j$, i.e. if i, j are in the same coplactic class, then [6.5.2(i)] says that i, j are in the same connected component of Γ, which means that we connect i and j by a chain of links, each link being of the form either $\xrightarrow{c}$ or $\xleftarrow{c}$. But this is "only if" for Theorem A.

Proof of Proposition B. Theorem [6.5.1(ii)] says that LLT operators are compatible with the equivalence $\equiv$. For example, if $i, j \in A^*$ and if $i \equiv j$, then for any $c \in \{1, \ldots, n-1\}$ there holds $f_c(i) \equiv f_c(j)$. (This includes the statement $f_c(i) = 0$ if and only if $f_c(j) = 0$.) But this is essentially the Lemma (C.6c). To deduce Proposition B, we combine (C.6c) with (C.3p), which says that for any i there holds $i \equiv KP(i)$. However LLT have proved this in Proposition [6.2.3]. Therefore LLT have proved Proposition B.

(D.11i) We sketch the proof that the LLT operators e_c, f_c are the same as the Littelmann operators $\tilde{e}_c$, $\tilde{f}_c$, respectively. Let $c \in \{1, \ldots, n-1\}$, and keep this fixed. To calculate $\tilde{e}_c(i)$ and $\tilde{f}_c(i)$ for a given word i of length p, use the function $h_c^i(t)$ (see (A.3b)). This gives parameters M^i, q^i, $\overline{q}^i$, and these are sufficient to determine $\tilde{e}_c(i)$ and $\tilde{f}_c(i)$. Let us say that words i, j are *isologous* if M^i, q^i, $\overline{q}^i$ are equal to M^j, q^j, $\overline{q}^j$, respectively.

To calculate $h_c^i(t)$, one needs only the entries c, $c+1$ in i. We say that letters other than c, $c+1$ are *neutral.* In the example below $i = 235342233$ is a word in $I(5,9)$, and $c = 2$. Our first move is to replace each neutral entry by the empty square, indicated by a "$\cdot$" in the third line of table D.1 below. Now we look for an *adjacent* $(c+1, c)$ *pair*, i.e. entries i_a, i_b of i

t	1	2	3	4	5	6	7	8	9
i_t	2	3	5	3	4	2	2	3	3
i_t	2	3	·	3	·	2	2	3	3
j_t	2	3	·	·	·	·	2	3	3
k_t	2	·	·	·	·	·	·	3	3

Table D.1. Successive construction of isologous words.

such that $i_a = c+1$, $i_b = c$, $a < b$, and i_z is neutral for all places z such that $a < z < b$, if there is any such place. In our example, (i_4, i_6) is an adjacent $(3, 2)$ pair. Now replace both entries i_a, i_b by neutral letters. It is (very) easy to see that the resulting word j is isologous to i.

We next look for adjacent $(c+1, c)$ pairs in j; in the example, (j_2, j_7) is such a pair. Then "neutralize" this pair, etc. After a finite number of steps we reach a word k that contains no adjacent $(c+1, c)$ pair. In this word, there may be r entries c, and they all occur before any of the s entries $c+1$. (Either or both of r, s may be zero.) By construction k is isologous to i. But it is very simple to describe the function $h_c^k(t)$: starting from the left, it ascends by the r steps c, moves horizontally if there are some neutral entries between the last c and the first $c+1$, then descends by the s entries $c+1$. If $r = 0$ then $M^i = M^k = 0$, and if $s = 0$ we have $h_c^i(p) = h_c^k(p) = M^k$; if $r > 0$ then $q^i = q^k$ is the last place with entry c, and if $s > 0$ then $\overline{q}^i = \overline{q}^k$ is the place immediately before the first $c+1$. In the example given in table D.1, we have exactly one c at place 1 in k, and two $c+1$'s at places 8, 9, respectively; hence $q^i = 1$ and $\overline{q}^i = 7$. We now have all that is needed to construct $\tilde{e}_c(i)$ and $\tilde{f}_c(i)$.

We leave it to the reader to compare our construction of $\tilde{e}_c$, $\tilde{f}_c$ with LLT's construction of their operators e_c, f_c, and to show that the two constructions are identical.

E

Tables

E.1 Schensted's decomposition of $I(3,3)$

Let $n = 3$ and $r = 3$. For each $i \in I(3,3)$, the Q-symbol $Q = Q(i)$ of i then contains each of the numbers 1, 2, 3 exactly once. We write $I(Q) = I(Q, \approx)$ for all these tableaux Q. Then

$$\begin{aligned} I(3,3) = \quad & I_{(300)} \quad \dot\cup \quad I_{(210)} \quad \dot\cup \quad I_{(111)} \\ = \; & I\left(\begin{array}{|c|c|c|}\hline 1&2&3\\\hline\end{array}\right) \;\dot\cup\; I\left(\begin{array}{|c|c|}\hline 1&2\\\hline 3\\\cline{1-1}\end{array}\right) \;\dot\cup\; I\left(\begin{array}{|c|c|}\hline 1&3\\\hline 2\\\cline{1-1}\end{array}\right) \;\dot\cup\; I\left(\begin{array}{|c|}\hline 1\\\hline 2\\\hline 3\\\hline\end{array}\right). \end{aligned}$$

Table E.1 below contains, for each $\lambda \in \Lambda^+(3,3)$, the tableaux $\psi^{(\lambda)}$, $\mathsf{Q}^{(\lambda)}$ (see (C.2g) and (C.2h)), the tableaux T_λ, Z_λ, and the words i^λ, i_λ obtained from these (see (D.1d) and (D.1e)).

<table>
<tr><th>λ</th><th>$\psi^{(\lambda)}$</th><th>$\mathsf{Q}^{(\lambda)}$</th><th>T_λ</th><th>i^λ</th><th>Z_λ</th><th>i_λ</th></tr>
<tr><td>$(3,0,0)$</td><td>$\begin{array}{|c|c|c|}\hline 1&2&3\\\hline\end{array}$</td><td>$\begin{array}{|c|c|c|}\hline 1&2&3\\\hline\end{array}$</td><td>$\begin{array}{|c|c|c|}\hline 1&1&1\\\hline\end{array}$</td><td>1 1 1</td><td>$\begin{array}{|c|c|c|}\hline 3&3&3\\\hline\end{array}$</td><td>3 3 3</td></tr>
<tr><td>$(2,1,0)$</td><td>$\begin{array}{|c|c|}\hline 2&3\\\hline 1\\\cline{1-1}\end{array}$</td><td>$\begin{array}{|c|c|}\hline 1&3\\\hline 2\\\cline{1-1}\end{array}$</td><td>$\begin{array}{|c|c|}\hline 1&1\\\hline 2\\\cline{1-1}\end{array}$</td><td>2 1 1</td><td>$\begin{array}{|c|c|}\hline 2&3\\\hline 3\\\cline{1-1}\end{array}$</td><td>3 2 3</td></tr>
<tr><td>$(1,1,1)$</td><td>$\begin{array}{|c|}\hline 3\\\hline 2\\\hline 1\\\hline\end{array}$</td><td>$\begin{array}{|c|}\hline 1\\\hline 2\\\hline 3\\\hline\end{array}$</td><td>$\begin{array}{|c|}\hline 1\\\hline 2\\\hline 3\\\hline\end{array}$</td><td>3 2 1</td><td>$\begin{array}{|c|}\hline 1\\\hline 2\\\hline 3\\\hline\end{array}$</td><td>3 2 1</td></tr>
</table>

Table E.1. Various data associated with $\lambda \in \Lambda^+(3,3)$.

The elements of the sets $I(Q)$ with their P-symbols and Q-symbols are listed in table E.2.

$\lambda = (3,0,0)$

$P(i)$	i
1 1 1	1 1 1
1 1 2	1 1 2
1 2 2	1 2 2
1 1 3	1 1 3
2 2 2	2 2 2
1 2 3	1 2 3
2 2 3	2 2 3
1 3 3	1 3 3
2 3 3	2 3 3
3 3 3	3 3 3
$Q(i)$	1 2 3

$\lambda = (2,1,0)$

$P(i)$	i	
1 1 / 2	1 2 1	2 1 1
1 2 / 2	2 2 1	2 1 2
1 1 / 3	1 3 1	3 1 1
1 3 / 2	2 3 1	2 1 3
1 2 / 3	1 3 2	3 1 2
1 3 / 3	3 3 1	3 1 3
2 2 / 3	2 3 2	3 2 2
2 3 / 3	3 3 2	3 2 3
$Q(i)$	1 2 / 3	1 3 / 2

$\lambda = (1,1,1)$

$P(i)$	i
1 / 2 / 3	3 2 1
$Q(i)$	1 / 2 / 3

Table E.2. P-symbols and Q-symbols of the words $i \in I(3,3)$.

E.2 The Littelmann graph $I(3,3)$

Let n, r be positive integers. Following Littelmann [35, §2], we define the structure of a graph on $I(n,r)$ by saying that $i, j \in I(n,r)$ are connected by an edge if there exists an element $c \in \{1, \dots, n-1\}$ such that $\tilde{f}_c(i) = j$ or $\tilde{f}_c(j) = i$.

This graph is the *Littelmann graph* (it is the undirected form of the directed graph Γ in (D.11i)). The connected components of this graph are precisely the *coplactic*, or $\approx$*-classes* $I(Q, \approx)$, where Q is a standard tableau with entries $1, \dots, r$ in some order. This follows from Theorem A (and (A.3g)(5), where we have seen that $\tilde{f}_c(i) = j$ if and only if $\tilde{e}_c(j) = i$). Therefore we can use these tableaux to label the connected components of the Littelmann graph.

For $n = r = 3$, the Littelmann graph is shown in table E.3. In this display, two words i, j are connected by a (directed) edge labelled by 1 or 2 according as $\tilde{f}_1(i) = j$ or $\tilde{f}_2(i) = j$.

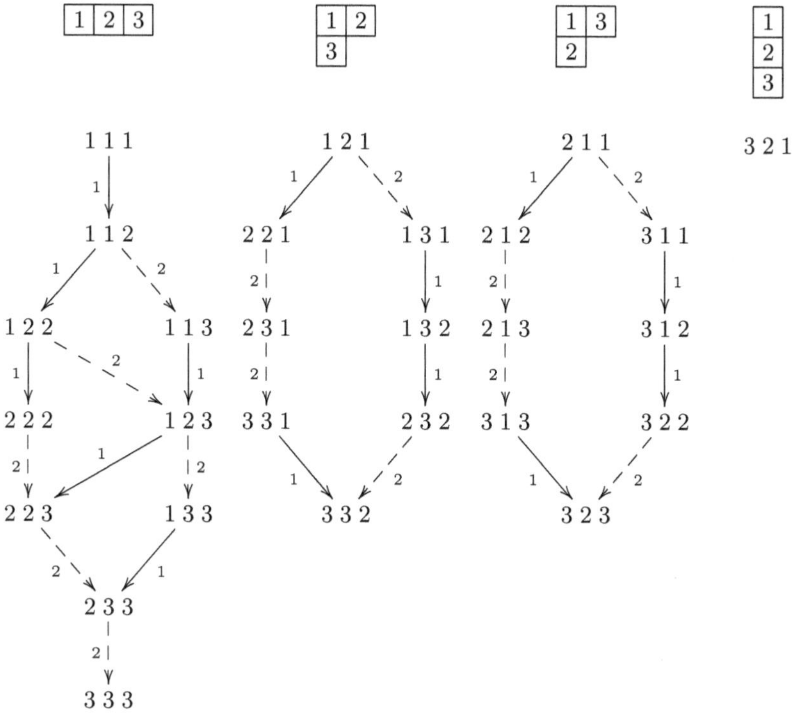

Table E.3. The four Littelmann graphs in $I(3,3)$.

Note that, if $\lambda \in \Lambda^+(n,r)$ and $Q = \mathsf{Q}^{(\lambda)}$, then i^λ is at the top and i_λ is at the bottom of the corresponding connected component of the Littelmann graph (see table E.1 and (D.2d), (C.2c)).

Index of symbols

References

1. E. Artin, C. Nesbitt, and R. M. Thrall. *Rings with minimum condition.* University of Michigan Press, Ann Arbor, 1948.
2. M. Auslander. Representation theory of Artin algebras I, II. *Comm. in Alg.*, 1:177–268, 1974.
3. D. Blessenohl and M. Schocker. *Noncommutative representation theory of the symmetric group.* Imperial College Press, London, 2005.
4. A. Borel. Properties and representations of Chevalley groups. In *Seminar on Algebraic Groups and Related Finite Groups (The Institute for Advanced Study, Princeton, N.J.), Lecture Notes in Mathematics*, volume 131, pages 1–55. Springer, Berlin, 1971.
5. R. Brauer. On modular and p-adic representations of algebras. *Proc. Nat. Acad. Sci. U.S.A.*, 25:252–258, 1939.
6. R.W. Carter and G. Lusztig. On the modular representations of the general linear and symmetric groups. *Math. Z.*, 136:193–242, 1974.
7. C. Chevalley. Certains schémas de groupes semi-simples. *Séminaire Bourbaki, Soc. Math. France, Paris,* Exp. No. 219, Vol. 6:219–234, 1995.
8. M. Clausen. Letter place algebras and a characteristic-free approach to the representation theory of the general linear and symmetric groups, I. *Adv. in Math.*, 33:161–191, 1979.
9. E. Cline, B. Parshall, and L. Scott. Cohomology, hyperalgebras, and representations. *J. Algebra*, 63:98–123, 1980.
10. P. M. Cohn. *Morita equivalence and duality.* Queen Mary College Mathematics Notes, University of London, 1976.
11. C. W. Curtis and I. Reiner. *Representation theory of finite groups and associative algebras.* John Wiley and Sons (Interscience), New York, 1962.
12. C.W. Curtis and T. V. Fossum. On centralizer rings and characters of representations of finite groups. *Math. Z.*, 107:402–406, 1968.
13. J. Deruyts. Essai d'une théorie générale des formes algébriques. *Mém. Soc. Roy. Liège*, 17:1–156, 1892.
14. J. Désarménien. *appendix to "Théorie combinatoire des invariants classiques"*, volume 1/S-01 of *Séries de Math. pures et appl.* G.-C. Rota, Université Louis-Pasteur, Strasbourg, 1977.
15. J. Désarménien, J. P. S. Kung, and G.-C. Rota. Invariant theory, Young bitableaux and combinatorics. *Adv. in Math.*, 27:63–92, 1978.

16. L. Dornhoff. *Group representation theory.* Marcel Dekker, New York, 1972.
17. F. G. Frobenius. Über die Charaktere der symmetrischen Gruppe. *Sitzber. Kgl. Preuß. Akad. Wiss.*, pages 516–534, 1900.
18. W. Fulton. *Young Tableaux*, volume 35 of *London Mathematical Society, Student Texts.* Cambridge University Press, 1997.
19. H. Garnir. *Théorie de la représentation linéaire des group symétriques*, volume 10 of *Mém. Soc. Roy. Sci. (4).* Liège, 1950.
20. J. A. Green. Locally finite representations. *J. Algebra*, 41:137–171, 1976.
21. C. Greene. An extension of Schensted's theorem. *Adv. in Math.*, 14:254–265, 1974.
22. W. J. Haboush. Central differential operators on split semi-simple groups over fields of positive characteristic. In *Séminaire d'Algebre P. Dubreil et Marie-Paule, Springer Lecture Notes in Math. 795*, pages 35–85. Springer, Berlin, 1980.
23. G. Higman. Representations of general linear groups and varieties of p-groups. In *Proc. Internat. Conf. Theory of Groups, Austral. Nat. Univ. Canberra, Aug. 1965*, pages 167–173. Gordon and Breach, New York, 1967.
24. G. Hochschild. *Introduction to affine algebraic groups.* Holden-Day, San Francisco, 1971.
25. N. Iwahori. On the structure of a Hecke ring of a Chevalley group over a finite field. *J. Fac. Sci. Univ. Tokyo Sect. I*, 10:215–236, 1964.
26. G. D. James. Some counterexamples in the theory of Specht modules. *J. Algebra*, 46:457–461, 1977.
27. G. D. James. *The representation theory of the symmetric group*, volume 682 of *Lecture Notes in Mathematics.* Springer, Berlin, 1978.
28. G. D. James. The decomposition of tensors over fields of prime characteristic. *Math. Z.*, 172:161–178, 1980.
29. J. C. Jantzen. Darstellungen halbeinfacher algebraischer Gruppen und zugeordnete kontravariante Formen. In *Bonner Math. Schriften*, volume 67. Bonn, 1973.
30. J. C. Jantzen. Darstellung halbeinfacher Gruppen und kontravariante Formen. *J. reine angew. Math.*, 290:117–141, 1977.
31. J. C. Jantzen. Über das Dekompositionsverhalten gewisser modularer Darstellungen halbeinfacher Gruppen und ihrer Lie-Algebren. *J. Algebra*, 49:441–469, 1977.
32. J. C. Jantzen. Weyl modules for groups of Lie type. In *Proc. of London Math. Soc. Symposium in Finite Simple Groups.* Durham, 1978.
33. V. Kac. *Infinite dimensional Lie algebras, Third edition.* Cambridge University Press, Cambridge, 1990.
34. D. E. Knuth. Permutations, matrices and generalized Young tableaux. *Pacific J. Math.*, 34:709–727, 1970.
35. P. Littelmann. Paths and root operators in representation theory. *Annals of Math.*, 142:499–525, 1995.
36. D. E. Littlewood. *The theory of group characters.* Oxford University Press (Clarendon), Oxford, 1950.
37. D. E. Littlewood and R. Richardson. Group Characters and Algebra. *Philos. Trans. of the Royal Soc. of Lond., Ser. A*, 233:99–141, 1934.
38. M. Lothaire. *Algebraic combinatorics on words*, volume 90 of *Encyclopedia of Mathematics and its Applications.* Cambridge University Press, Cambridge, 2002.

39. I. G. Macdonald. *Symmetric functions and Hall polynomials.* Oxford University Press (Clarendon), Oxford, 1979.
40. P. A. MacMahon. *Combinatory Analysis I, II.* Cambridge University Press, 1915/16, reprint by Chelsea Publishing Company, 1960.
41. T. Martins. Hook representations of the general linear group. Ph.D. thesis, Warwick University, Coventry, 1981.
42. T. Nakayama. On Frobeniusean algebras I. *Ann. of Math.*, 40:611–633, 1939.
43. T. Nakayama. On Frobeniusean algebras II. *Ann. of Math.*, 42:1–21, 1941.
44. T. Nakayama. On Frobeniusean algebras III. *Jap. J. Math.*, 18:49–65, 1942.
45. M. H. Peel. Specht modules and symmetric groups. *J. Algebra*, 36:88–97, 1975.
46. C. Schensted. Longest increasing and decreasing subsequences. *Canad. J. Math.*, 13:179–191, 1961.
47. I. Schur. Über eine Klasse von Matrizen, die sich einer gegebenen Matrix zuordnen lassen. Dissertation, Berlin, 1901. In I. Schur, Gesammelte Abhandlungen I, 1–70, Springer, Berlin, 1973.
48. I. Schur. Über die rationalen Darstellungen der allgemeinen linearen Gruppe. *Sitzber. Königl. Preuß. Ak. Wiss., Physikal.-Math. Klasse*, pages 58–75, 1927. In I. Schur, Gesammelte Abhandlungen III, 68–85, Springer, Berlin, 1973.
49. J.-P. Serre. Groupes de Grothendieck des schémas en groupes réductifs déployés. *Publ. I.H.E.S.*, 34:37–52, 1968.
50. R. Steinberg. *Lectures on Chevalley groups.* Yale University, New Haven, 1966.
51. M. E. Sweedler. *Hopf algebras.* W. A. Benjamin, New York, 1969.
52. J. Towber. Young symmetry, the flag manifold, and representations of GL(n). *J. Algebra*, 61:414–462, 1979.
53. D.-N. Verma. The role of affine Weyl groups in the representation theory of algebraic Chevalley groups and their Lie algebra. In I. M. Gelfand, editor, *Lie groups and their representations*, pages 653–705. John Wiley and Sons, New York, 1975.
54. H. Weyl. Theorie der Darstellung kontinuierlicher halbeinfacher Gruppen durch lineare Transformationen. *Math. Z.*, 23:271–309, 1925. 24:328–376, 377–395, 789–791, 1926.
55. H. Weyl. *The classical groups.* Princeton University Press, Princeton, 1946.
56. W. J. Wong. Representations of Chevalley groups in characteristic p. *Nagoya Math. J.*, 45:39–78, 1971.
57. W. J. Wong. Irreducible modular representations of finite Chevalley groups. *J. Algebra*, 20:355–367, 1972.
58. A. Young. On quantitative substitutional analysis (2). *Proc. London Math. Soc.*, 34(1):261–397, 1902.
59. A. Young. On quantitative substitutional analysis (3). *Proc. London Math. Soc.*, 28(2):255–292, 1928.

38. I. G. Macdonald, *Symmetric functions and Hall polynomials*, Oxford University Press (Clarendon), Oxford, 1979.
39. P. A. MacMahon, *Combinatory Analysis*, I, II, Cambridge University Press, 1915-16, reprint by Chelsea Publishing Company, 1960.
40. [illegible], Hook representations of the general linear group, PhD thesis, Warwick University, Coventry, 1981.
41. [illegible]
42. T. Nakayama, [illegible], *Ann. of Math.* [illegible]
43. T. Nakayama, On Frobeniusean algebras [illegible], *Ann. of Math.* 42, 1-21, 1941.
44. T. Nakayama, [illegible]
45. [illegible]
46. C. Schensted, Longest increasing and decreasing subsequences, *Canad. J. Math.* 13, 179-191, 1961.
47. I. Schur, Über eine Klasse von Matrizen, die sich einer gegebenen Matrix zuordnen lassen, Dissertation, Berlin 1901. In I. Schur, *Gesammelte Abhandlungen* I, 1-72, Springer, Berlin 1973.
48. I. Schur, Über die rationalen Darstellungen der allgemeinen linearen Gruppe, *Sitzber. Königl. Preuß. Ak. Wiss., Physikal.-Math. Klasse*, pages 58-75, 1927. In I. Schur, *Gesammelte Abhandlungen* III, 68-85, Springer, Berlin, 1973.
49. [illegible]
50. R. Steinberg, *Lectures on Chevalley groups*, Yale University, New Haven, 1968.
51. [illegible], New York, 1968.
52. [illegible]
53. [illegible]
54. [illegible]
55. H. Weyl, *The classical groups*, Princeton University Press, Princeton, 1946.
56. [illegible]
57. [illegible]
58. A. Young, On quantitative substitutional analysis [illegible]
59. [illegible]

Index

Lecture Notes in Mathematics

For information about earlier volumes
please contact your bookseller or Springer
LNM Online archive: springerlink.com

Vol. 1711: W. Ricker, Operator Algebras Generated by Commuting Projećtions: A Vector Measure Approach (1999)

Vol. 1712: N. Schwartz, J. J. Madden, Semi-algebraic Function Rings and Reflectors of Partially Ordered Rings (1999)

Vol. 1713: F. Bethuel, G. Huisken, S. Müller, K. Steffen, Calculus of Variations and Geometric Evolution Problems. Cetraro, 1996. Editors: S. Hildebrandt, M. Struwe (1999)

Vol. 1714: O. Diekmann, R. Durrett, K. P. Hadeler, P. K. Maini, H. L. Smith, Mathematics Inspired by Biology. Martina Franca, 1997. Editors: V. Capasso, O. Diekmann (1999)

Vol. 1715: N. V. Krylov, M. Röckner, J. Zabczyk, Stochastic PDE's and Kolmogorov Equations in Infinite Dimensions. Cetraro, 1998. Editor: G. Da Prato (1999)

Vol. 1716: J. Coates, R. Greenberg, K. A. Ribet, K. Rubin, Arithmetic Theory of Elliptic Curves. Cetraro, 1997. Editor: C. Viola (1999)

Vol. 1717: J. Bertoin, F. Martinelli, Y. Peres, Lectures on Probability Theory and Statistics. Saint-Flour, 1997. Editor: P. Bernard (1999)

Vol. 1718: A. Eberle, Uniqueness and Non-Uniqueness of Semigroups Generated by Singular Diffusion Operators (1999)

Vol. 1719: K. R. Meyer, Periodic Solutions of the N-Body Problem (1999)

Vol. 1720: D. Elworthy, Y. Le Jan, X-M. Li, On the Geometry of Diffusion Operators and Stochastic Flows (1999)

Vol. 1721: A. Iarrobino, V. Kanev, Power Sums, Gorenstein Algebras, and Determinantal Loci (1999)

Vol. 1722: R. McCutcheon, Elemental Methods in Ergodic Ramsey Theory (1999)

Vol. 1723: J. P. Croisille, C. Lebeau, Diffraction by an Immersed Elastic Wedge (1999)

Vol. 1724: V. N. Kolokoltsov, Semiclassical Analysis for Diffusions and Stochastic Processes (2000)

Vol. 1725: D. A. Wolf-Gladrow, Lattice-Gas Cellular Automata and Lattice Boltzmann Models (2000)

Vol. 1726: V. Marić, Regular Variation and Differential Equations (2000)

Vol. 1727: P. Kravanja M. Van Barel, Computing the Zeros of Analytic Functions (2000)

Vol. 1728: K. Gatermann Computer Algebra Methods for Equivariant Dynamical Systems (2000)

Vol. 1729: J. Azéma, M. Émery, M. Ledoux, M. Yor (Eds.) Séminaire de Probabilités XXXIV (2000)

Vol. 1730: S. Graf, H. Luschgy, Foundations of Quantization for Probability Distributions (2000)

Vol. 1731: T. Hsu, Quilts: Central Extensions, Braid Actions, and Finite Groups (2000)

Vol. 1732: K. Keller, Invariant Factors, Julia Equivalences and the (Abstract) Mandelbrot Set (2000)

Vol. 1733: K. Ritter, Average-Case Analysis of Numerical Problems (2000)

Vol. 1734: M. Espedal, A. Fasano, A. Mikelić, Filtration in Porous Media and Industrial Applications. Cetraro 1998. Editor: A. Fasano. 2000.

Vol. 1735: D. Yafaev, Scattering Theory: Some Old and New Problems (2000)

Vol. 1736: B. O. Turesson, Nonlinear Potential Theory and Weighted Sobolev Spaces (2000)

Vol. 1737: S. Wakabayashi, Classical Microlocal Analysis in the Space of Hyperfunctions (2000)

Vol. 1738: M. Émery, A. Nemirovski, D. Voiculescu, Lectures on Probability Theory and Statistics (2000)

Vol. 1739: R. Burkard, P. Deuflhard, A. Jameson, J.-L. Lions, G. Strang, Computational Mathematics Driven by Industrial Problems. Martina Franca, 1999. Editors: V. Capasso, H. Engl, J. Periaux (2000)

Vol. 1740: B. Kawohl, O. Pironneau, L. Tartar, J.-P. Zolesio, Optimal Shape Design. Tróia, Portugal 1999. Editors: A. Cellina, A. Ornelas (2000)

Vol. 1741: E. Lombardi, Oscillatory Integrals and Phenomena Beyond all Algebraic Orders (2000)

Vol. 1742: A. Unterberger, Quantization and Non-holomorphic Modular Forms (2000)

Vol. 1743: L. Habermann, Riemannian Metrics of Constant Mass and Moduli Spaces of Conformal Structures (2000)

Vol. 1744: M. Kunze, Non-Smooth Dynamical Systems (2000)

Vol. 1745: V. D. Milman, G. Schechtman (Eds.), Geometric Aspects of Functional Analysis. Israel Seminar 1999-2000 (2000)

Vol. 1746: A. Degtyarev, I. Itenberg, V. Kharlamov, Real Enriques Surfaces (2000)

Vol. 1747: L. W. Christensen, Gorenstein Dimensions (2000)

Vol. 1748: M. Ruzicka, Electrorheological Fluids: Modeling and Mathematical Theory (2001)

Vol. 1749: M. Fuchs, G. Seregin, Variational Methods for Problems from Plasticity Theory and for Generalized Newtonian Fluids (2001)

Vol. 1750: B. Conrad, Grothendieck Duality and Base Change (2001)

Vol. 1751: N. J. Cutland, Loeb Measures in Practice: Recent Advances (2001)

Vol. 1752: Y. V. Nesterenko, P. Philippon, Introduction to Algebraic Independence Theory (2001)

Vol. 1753: A. I. Bobenko, U. Eitner, Painlevé Equations in the Differential Geometry of Surfaces (2001)

Vol. 1754: W. Bertram, The Geometry of Jordan and Lie Structures (2001)

Vol. 1755: J. Azéma, M. Émery, M. Ledoux, M. Yor (Eds.), Séminaire de Probabilités XXXV (2001)

Vol. 1756: P. E. Zhidkov, Korteweg de Vries and Nonlinear Schrödinger Equations: Qualitative Theory (2001)

Vol. 1757: R. R. Phelps, Lectures on Choquet's Theorem (2001)
Vol. 1758: N. Monod, Continuous Bounded Cohomology of Locally Compact Groups (2001)
Vol. 1759: Y. Abe, K. Kopfermann, Toroidal Groups (2001)
Vol. 1760: D. Filipović, Consistency Problems for Heath-Jarrow-Morton Interest Rate Models (2001)
Vol. 1761: C. Adelmann, The Decomposition of Primes in Torsion Point Fields (2001)
Vol. 1762: S. Cerrai, Second Order PDE's in Finite and Infinite Dimension (2001)
Vol. 1763: J.-L. Loday, A. Frabetti, F. Chapoton, F. Goichot, Dialgebras and Related Operads (2001)
Vol. 1764: A. Cannas da Silva, Lectures on Symplectic Geometry (2001)
Vol. 1765: T. Kerler, V. V. Lyubashenko, Non-Semisimple Topological Quantum Field Theories for 3-Manifolds with Corners (2001)
Vol. 1766: H. Hennion, L. Hervé, Limit Theorems for Markov Chains and Stochastic Properties of Dynamical Systems by Quasi-Compactness (2001)
Vol. 1767: J. Xiao, Holomorphic Q Classes (2001)
Vol. 1768: M.J. Pflaum, Analytic and Geometric Study of Stratified Spaces (2001)
Vol. 1769: M. Alberich-Carramiñana, Geometry of the Plane Cremona Maps (2002)
Vol. 1770: H. Gluesing-Luerssen, Linear Delay-Differential Systems with Commensurate Delays: An Algebraic Approach (2002)
Vol. 1771: M. Émery, M. Yor (Eds.), Séminaire de Probabilités 1967-1980. A Selection in Martingale Theory (2002)
Vol. 1772: F. Burstall, D. Ferus, K. Leschke, F. Pedit, U. Pinkall, Conformal Geometry of Surfaces in S^4 (2002)
Vol. 1773: Z. Arad, M. Muzychuk, Standard Integral Table Algebras Generated by a Non-real Element of Small Degree (2002)
Vol. 1774: V. Runde, Lectures on Amenability (2002)
Vol. 1775: W. H. Meeks, A. Ros, H. Rosenberg, The Global Theory of Minimal Surfaces in Flat Spaces. Martina Franca 1999. Editor: G. P. Pirola (2002)
Vol. 1776: K. Behrend, C. Gomez, V. Tarasov, G. Tian, Quantum Comohology. Cetraro 1997. Editors: P. de Bartolomeis, B. Dubrovin, C. Reina (2002)
Vol. 1777: E. García-Río, D. N. Kupeli, R. Vázquez-Lorenzo, Osserman Manifolds in Semi-Riemannian Geometry (2002)
Vol. 1778: H. Kiechle, Theory of K-Loops (2002)
Vol. 1779: I. Chueshov, Monotone Random Systems (2002)
Vol. 1780: J. H. Bruinier, Borcherds Products on O(2,1) and Chern Classes of Heegner Divisors (2002)
Vol. 1781: E. Bolthausen, E. Perkins, A. van der Vaart, Lectures on Probability Theory and Statistics. Ecole d' Eté de Probabilités de Saint-Flour XXIX-1999. Editor: P. Bernard (2002)
Vol. 1782: C.-H. Chu, A. T.-M. Lau, Harmonic Functions on Groups and Fourier Algebras (2002)
Vol. 1783: L. Grüne, Asymptotic Behavior of Dynamical and Control Systems under Perturbation and Discretization (2002)
Vol. 1784: L.H. Eliasson, S. B. Kuksin, S. Marmi, J.-C. Yoccoz, Dynamical Systems and Small Divisors. Cetraro, Italy 1998. Editors: S. Marmi, J.-C. Yoccoz (2002)
Vol. 1785: J. Arias de Reyna, Pointwise Convergence of Fourier Series (2002)
Vol. 1786: S. D. Cutkosky, Monomialization of Morphisms from 3-Folds to Surfaces (2002)
Vol. 1787: S. Caenepeel, G. Militaru, S. Zhu, Frobenius and Separable Functors for Generalized Module Categories and Nonlinear Equations (2002)
Vol. 1788: A. Vasil'ev, Moduli of Families of Curves for Conformal and Quasiconformal Mappings (2002)
Vol. 1789: Y. Sommerhäuser, Yetter-Drinfel'd Hopf algebras over groups of prime order (2002)
Vol. 1790: X. Zhan, Matrix Inequalities (2002)
Vol. 1791: M. Knebusch, D. Zhang, Manis Valuations and Prüfer Extensions I: A new Chapter in Commutative Algebra (2002)
Vol. 1792: D. D. Ang, R. Gorenflo, V. K. Le, D. D. Trong, Moment Theory and Some Inverse Problems in Potential Theory and Heat Conduction (2002)
Vol. 1793: J. Cortés Monforte, Geometric, Control and Numerical Aspects of Nonholonomic Systems (2002)
Vol. 1794: N. Pytheas Fogg, Substitution in Dynamics, Arithmetics and Combinatorics. Editors: V. Berthé, S. Ferenczi, C. Mauduit, A. Siegel (2002)
Vol. 1795: H. Li, Filtered-Graded Transfer in Using Noncommutative Gröbner Bases (2002)
Vol. 1796: J.M. Melenk, hp-Finite Element Methods for Singular Perturbations (2002)
Vol. 1797: B. Schmidt, Characters and Cyclotomic Fields in Finite Geometry (2002)
Vol. 1798: W.M. Oliva, Geometric Mechanics (2002)
Vol. 1799: H. Pajot, Analytic Capacity, Rectifiability, Menger Curvature and the Cauchy Integral (2002)
Vol. 1800: O. Gabber, L. Ramero, Almost Ring Theory (2003)
Vol. 1801: J. Azéma, M. Émery, M. Ledoux, M. Yor (Eds.), Séminaire de Probabilités XXXVI (2003)
Vol. 1802: V. Capasso, E. Merzbach, B.G. Ivanoff, M. Dozzi, R. Dalang, T. Mountford, Topics in Spatial Stochastic Processes. Martina Franca, Italy 2001. Editor: E. Merzbach (2003)
Vol. 1803: G. Dolzmann, Variational Methods for Crystalline Microstructure – Analysis and Computation (2003)
Vol. 1804: I. Cherednik, Ya. Markov, R. Howe, G. Lusztig, Iwahori-Hecke Algebras and their Representation Theory. Martina Franca, Italy 1999. Editors: V. Baldoni, D. Barbasch (2003)
Vol. 1805: F. Cao, Geometric Curve Evolution and Image Processing (2003)
Vol. 1806: H. Broer, I. Hoveijn. G. Lunther, G. Vegter, Bifurcations in Hamiltonian Systems. Computing Singularities by Gröbner Bases (2003)
Vol. 1807: V. D. Milman, G. Schechtman (Eds.), Geometric Aspects of Functional Analysis. Israel Seminar 2000-2002 (2003)
Vol. 1808: W. Schindler, Measures with Symmetry Properties (2003)
Vol. 1809: O. Steinbach, Stability Estimates for Hybrid Coupled Domain Decomposition Methods (2003)
Vol. 1810: J. Wengenroth, Derived Functors in Functional Analysis (2003)
Vol. 1811: J. Stevens, Deformations of Singularities (2003)
Vol. 1812: L. Ambrosio, K. Deckelnick, G. Dziuk, M. Mimura, V. A. Solonnikov, H. M. Soner, Mathematical Aspects of Evolving Interfaces. Madeira, Funchal, Portugal 2000. Editors: P. Colli, J. F. Rodrigues (2003)
Vol. 1813: L. Ambrosio, L. A. Caffarelli, Y. Brenier, G. Buttazzo, C. Villani, Optimal Transportation and its

Applications. Martina Franca, Italy 2001. Editors: L. A. Caffarelli, S. Salsa (2003)

Vol. 1814: P. Bank, F. Baudoin, H. Föllmer, L.C.G. Rogers, M. Soner, N. Touzi, Paris-Princeton Lectures on Mathematical Finance 2002 (2003)

Vol. 1815: A. M. Vershik (Ed.), Asymptotic Combinatorics with Applications to Mathematical Physics. St. Petersburg, Russia 2001 (2003)

Vol. 1816: S. Albeverio, W. Schachermayer, M. Talagrand, Lectures on Probability Theory and Statistics. Ecole d'Eté de Probabilités de Saint-Flour XXX-2000. Editor: P. Bernard (2003)

Vol. 1817: E. Koelink, W. Van Assche(Eds.), Orthogonal Polynomials and Special Functions. Leuven 2002 (2003)

Vol. 1818: M. Bildhauer, Convex Variational Problems with Linear, nearly Linear and/or Anisotropic Growth Conditions (2003)

Vol. 1819: D. Masser, Yu. V. Nesterenko, H. P. Schlickewei, W. M. Schmidt, M. Waldschmidt, Diophantine Approximation. Cetraro, Italy 2000. Editors: F. Amoroso, U. Zannier (2003)

Vol. 1820: F. Hiai, H. Kosaki, Means of Hilbert Space Operators (2003)

Vol. 1821: S. Teufel, Adiabatic Perturbation Theory in Quantum Dynamics (2003)

Vol. 1822: S.-N. Chow, R. Conti, R. Johnson, J. Mallet-Paret, R. Nussbaum, Dynamical Systems. Cetraro, Italy 2000. Editors: J. W. Macki, P. Zecca (2003)

Vol. 1823: A. M. Anile, W. Allegretto, C. Ringhofer, Mathematical Problems in Semiconductor Physics. Cetraro, Italy 1998. Editor: A. M. Anile (2003)

Vol. 1824: J. A. Navarro González, J. B. Sancho de Salas, $\mathscr{C}^\infty$ – Differentiable Spaces (2003)

Vol. 1825: J. H. Bramble, A. Cohen, W. Dahmen, Multiscale Problems and Methods in Numerical Simulations, Martina Franca, Italy 2001. Editor: C. Canuto (2003)

Vol. 1826: K. Dohmen, Improved Bonferroni Inequalities via Abstract Tubes. Inequalities and Identities of Inclusion-Exclusion Type. VIII, 113 p, 2003.

Vol. 1827: K. M. Pilgrim, Combinations of Complex Dynamical Systems. IX, 118 p, 2003.

Vol. 1828: D. J. Green, Gröbner Bases and the Computation of Group Cohomology. XII, 138 p, 2003.

Vol. 1829: E. Altman, B. Gaujal, A. Hordijk, Discrete-Event Control of Stochastic Networks: Multimodularity and Regularity. XIV, 313 p, 2003.

Vol. 1830: M. I. Gil', Operator Functions and Localization of Spectra. XIV, 256 p, 2003.

Vol. 1831: A. Connes, J. Cuntz, E. Guentner, N. Higson, J. E. Kaminker, Noncommutative Geometry, Martina Franca, Italy 2002. Editors: S. Doplicher, L. Longo (2004)

Vol. 1832: J. Azéma, M. Émery, M. Ledoux, M. Yor (Eds.), Séminaire de Probabilités XXXVII (2003)

Vol. 1833: D.-Q. Jiang, M. Qian, M.-P. Qian, Mathematical Theory of Nonequilibrium Steady States. On the Frontier of Probability and Dynamical Systems. IX, 280 p, 2004.

Vol. 1834: Yo. Yomdin, G. Comte, Tame Geometry with Application in Smooth Analysis. VIII, 186 p, 2004.

Vol. 1835: O.T. Izhboldin, B. Kahn, N.A. Karpenko, A. Vishik, Geometric Methods in the Algebraic Theory of Quadratic Forms. Summer School, Lens, 2000. Editor: J.-P. Tignol (2004)

Vol. 1836: C. Năstăsescu, F. Van Oystaeyen, Methods of Graded Rings. XIII, 304 p, 2004.

Vol. 1837: S. Tavaré, O. Zeitouni, Lectures on Probability Theory and Statistics. Ecole d'Eté de Probabilités de Saint-Flour XXXI-2001. Editor: J. Picard (2004)

Vol. 1838: A.J. Ganesh, N.W. O'Connell, D.J. Wischik, Big Queues. XII, 254 p, 2004.

Vol. 1839: R. Gohm, Noncommutative Stationary Processes. VIII, 170 p, 2004.

Vol. 1840: B. Tsirelson, W. Werner, Lectures on Probability Theory and Statistics. Ecole d'Eté de Probabilités de Saint-Flour XXXII-2002. Editor: J. Picard (2004)

Vol. 1841: W. Reichel, Uniqueness Theorems for Variational Problems by the Method of Transformation Groups (2004)

Vol. 1842: T. Johnsen, A.L. Knutsen, K3 Projective Models in Scrolls (2004)

Vol. 1843: B. Jefferies, Spectral Properties of Noncommuting Operators (2004)

Vol. 1844: K.F. Siburg, The Principle of Least Action in Geometry and Dynamics (2004)

Vol. 1845: Min Ho Lee, Mixed Automorphic Forms, Torus Bundles, and Jacobi Forms (2004)

Vol. 1846: H. Ammari, H. Kang, Reconstruction of Small Inhomogeneities from Boundary Measurements (2004)

Vol. 1847: T.R. Bielecki, T. Björk, M. Jeanblanc, M. Rutkowski, J.A. Scheinkman, W. Xiong, Paris-Princeton Lectures on Mathematical Finance 2003 (2004)

Vol. 1848: M. Abate, J. E. Fornaess, X. Huang, J. P. Rosay, A. Tumanov, Real Methods in Complex and CR Geometry, Martina Franca, Italy 2002. Editors: D. Zaitsev, G. Zampieri (2004)

Vol. 1849: Martin L. Brown, Heegner Modules and Elliptic Curves (2004)

Vol. 1850: V. D. Milman, G. Schechtman (Eds.), Geometric Aspects of Functional Analysis. Israel Seminar 2002-2003 (2004)

Vol. 1851: O. Catoni, Statistical Learning Theory and Stochastic Optimization (2004)

Vol. 1852: A.S. Kechris, B.D. Miller, Topics in Orbit Equivalence (2004)

Vol. 1853: Ch. Favre, M. Jonsson, The Valuative Tree (2004)

Vol. 1854: O. Saeki, Topology of Singular Fibers of Differential Maps (2004)

Vol. 1855: G. Da Prato, P.C. Kunstmann, I. Lasiecka, A. Lunardi, R. Schnaubelt, L. Weis, Functional Analytic Methods for Evolution Equations. Editors: M. Iannelli, R. Nagel, S. Piazzera (2004)

Vol. 1856: K. Back, T.R. Bielecki, C. Hipp, S. Peng, W. Schachermayer, Stochastic Methods in Finance, Bressanone/Brixen, Italy, 2003. Editors: M. Fritelli, W. Runggaldier (2004)

Vol. 1857: M. Émery, M. Ledoux, M. Yor (Eds.), Séminaire de Probabilités XXXVIII (2005)

Vol. 1858: A.S. Cherny, H.-J. Engelbert, Singular Stochastic Differential Equations (2005)

Vol. 1859: E. Letellier, Fourier Transforms of Invariant Functions on Finite Reductive Lie Algebras (2005)

Vol. 1860: A. Borisyuk, G.B. Ermentrout, A. Friedman, D. Terman, Tutorials in Mathematical Biosciences I. Mathematical Neurosciences (2005)

Vol. 1861: G. Benettin, J. Henrard, S. Kuksin, Hamiltonian Dynamics – Theory and Applications, Cetraro, Italy, 1999. Editor: A. Giorgilli (2005)

Vol. 1862: B. Helffer, F. Nier, Hypoelliptic Estimates and Spectral Theory for Fokker-Planck Operators and Witten Laplacians (2005)

Vol. 1863: H. Führh, Abstract Harmonic Analysis of Continuous Wavelet Transforms (2005)
Vol. 1864: K. Efstathiou, Metamorphoses of Hamiltonian Systems with Symmetries (2005)
Vol. 1865: D. Applebaum, B.V. R. Bhat, J. Kustermans, J. M. Lindsay, Quantum Independent Increment Processes I. From Classical Probability to Quantum Stochastic Calculus. Editors: M. Schürmann, U. Franz (2005)
Vol. 1866: O.E. Barndorff-Nielsen, U. Franz, R. Gohm, B. Kümmerer, S. Thorbjønsen, Quantum Independent Increment Processes II. Structure of Quantum Levy Processes, Classical Probability, and Physics. Editors: M. Schürmann, U. Franz, (2005)
Vol. 1867: J. Sneyd (Ed.), Tutorials in Mathematical Biosciences II. Mathematical Modeling of Calcium Dynamics and Signal Transduction. (2005)
Vol. 1868: J. Jorgenson, S. Lang, $Pos_n(R)$ and Eisenstein Sereies. (2005)
Vol. 1869: A. Dembo, T. Funaki, Lectures on Probability Theory and Statistics. Ecole d'Eté de Probabilités de Saint-Flour XXXIII-2003. Editor: J. Picard (2005)
Vol. 1870: V.I. Gurariy, W. Lusky, Geometry of Müntz Spaces and Related Questions. (2005)
Vol. 1871: P. Constantin, G. Gallavotti, A.V. Kazhikhov, Y. Meyer, S. Ukai, Mathematical Foundation of Turbulent Viscous Flows, Martina Franca, Italy, 2003. Editors: M. Cannone, T. Miyakawa (2006)
Vol. 1872: A. Friedman (Ed.), Tutorials in Mathematical Biosciences III. Cell Cycle, Proliferation, and Cancer (2006)
Vol. 1873: R. Mansuy, M. Yor, Random Times and Enlargements of Filtrations in a Brownian Setting (2006)
Vol. 1874: M. Yor, M. Émery (Eds.), In Memoriam Paul-André Meyer - Séminaire de Probabilités XXXIX (2006)
Vol. 1875: J. Pitman, Combinatorial Stochastic Processes. Ecole d'Eté de Probabilités de Saint-Flour XXXII-2002. Editor: J. Picard (2006)
Vol. 1876: H. Herrlich, Axiom of Choice (2006)
Vol. 1877: J. Steuding, Value Distributions of L-Functions(2006)
Vol. 1878: R. Cerf, The Wulff Crystal in Ising and Percolation Models, Ecole d'Eté de Probabilits de Saint-Flour XXXIV-2004. Editor: Jean Picard (2006)
Vol. 1879: G. Slade, The Lace Expansion and its Appli- cations, Ecole d'Eté de Probabilités de Saint-Flour XXXIV-2004. Editor: Jean Picard (2006)
Vol. 1880: S. Attal, A. Joye, C.-A. Pillet, Open Quantum Systems I, The Hamiltonian Approach (2006)
Vol. 1881: S. Attal, A. Joye, C.-A. Pillet, Open Quantum Systems II, The Markovian Approach (2006)
Vol. 1882: S. Attal, A. Joye, C.-A. Pillet, Open Quantum Systems III, Recent Developments (2006)
Vol. 1883: W. Van Assche, F. Marcellàn (Eds.), Orthogonal Polynomials and Special Functions, Computation and Application (2006)
Vol. 1884: N. Hayashi, E.I. Kaikina, P.I. Naumkin, I.A. Shishmarev, Asymptotics for Dissipative Nonlinear Equations (2006)
Vol. 1885: A. Telcs, The Art of Random Walks (2006)
Vol. 1886: S. Takamura, Splitting Deformations of Degenerations of Complex Curves (2006)
Vol. 1887: K. Habermann, L. Habermann, Introduction to Symplectic Dirac Operators (2006)
Vol. 1888: J. van der Hoeven, Transseries and Real Differential Algebra (2006)
Vol. 1889: G. Osipenko, Dynamical Systems, Graphs, and Algorithms (2006)
Vol. 1890: M. Bunge, J. Funk, Singular Coverings of Toposes (2006)
Vol. 1891: J.B. Friedlander, D.R. Heath-Brown, H. Iwaniec, J. Kaczorowski, Analytic Number Theory, Cetraro, Italy, 2002. Editors: A. Perelli, C. Viola (2006)
Vol. 1892: A. Baddeley, I. Bárány, R. Schneider, W. Weil, Stochastic Geometry, Martina Franca, Italy, 2004. Editor: W. Weil (2007)
Vol. 1893: H. Hanßmann, Local and Semi-Local Bifurcations in Hamiltonian Dynamical Systems, Results and Examples (2007)
Vol. 1894: C.W. Groetsch, Stable Approximate Evaluation of Unbounded Operators (2007)
Vol. 1895: L. Molnár, Selected Preserver Problems on Algebraic Structures of Linear Operators and on Function Spaces (2007)
Vol. 1896: P. Massart, Concentration Inequalities and Model Selection, Ecole d'Eté de Probabilitiés de Saint-Flour XXXIII-2003. Editor: J. Picard (2007)
Vol. 1897: R. Doney, Fluctuation Theory for Lévy Processes, Ecole d'Eté de Probabilitiés de Saint-Flour-2005. Editor: J. Picard (2007)
Vol. 1898: H.R. Beyer, Beyond Partial Differential Equations, On linear and Quasi-Linear Abstract Hyperbolic Evolution Equations (2007)
Vol. 1899: Seminaires de Probabilitiés XL. Editors: C. Donati-Martin, M. Émery, A. Rouault, C. Stricker (2007)
Vol. 1900: E. Bolthausen, A. Bovier (Eds.), Spin Glasses (2007)

Recent Reprints and New Editions

Vol. 1618: G. Pisier, Similarity Problems and Completely Bounded Maps. 1995 – 2nd exp. edition (2001)
Vol. 1629: J.D. Moore, Lectures on Seiberg-Witten Invariants. 1997 – 2nd edition (2001)
Vol. 1638: P. Vanhaecke, Integrable Systems in the realm of Algebraic Geometry. 1996 – 2nd edition (2001)
Vol. 1702: J. Ma, J. Yong, Forward-Backward Stochastic Differential Equations and their Applications. 1999. – Corr. 3rd printing (2005)
Vol. 830: J.A. Green, Polynomial Representations of GL_n, with an Appendix on Schensted Correspondence and Littelmann Paths by K. Erdmann, J.A. Green and M. Schocker. 1980 – 2nd corr. and augmented edition (2007)